W0017465

INNOVATION FOR DEVELOPMENT AND THE ROLE OF GOVERNMENT

A Perspective from the East Asia and Pacific Region

INNOVATION FOR DEVELOPMENT AND THE ROLE OF GOVERNMENT

A Perspective from the East Asia and Pacific Region

Edited by
Qimiao Fan
Kouqing Li
Douglas Zhihua Zeng
Yang Dong
Runzhong Peng

THE WORLD BANK
Washington, DC

©2009 The International Bank for Reconstruction and Development / The World Bank
1818 H Street, NW
Washington, DC 20433
Telephone: 202-473-1000
Internet: www.worldbank.org
E-mail: feedback

All rights reserved

1 2 3 4   11 10 09 08

This volume is a product of the staff of the International Bank for Reconstruction and Development / The World Bank. The findings, interpretations, and conclusions expressed in this volume do not necessarily reflect the views of the Executive Directors of The World Bank or the governments they represent.

The World Bank does not guarantee the accuracy of the data included in this work. The boundaries, colors, denominations, and other information shown on any map in this work do not imply any judgement on the part of The World Bank concerning the legal status of any territory or the endorsement or acceptance of such boundaries.

Rights and Permissions
The material in this publication is copyrighted. Copying and / or transmitting portions or all of this work without permission may be a violation of applicable law. The International Bank for Reconstruction and Development / The World Bank encourages dissemination of its work and will normally grant permission to reproduce portions of the work promptly.

For permission to photocopy or reprint any part of this work, please send a request with complete information to the Copyright Clearance Center Inc., 222 Rosewood Drive, Danvers, MA 01923, USA; telephone: 978-750-8400; fax: 978-750-4470; Internet: www.copyright.com.

All other queries on rights and licenses, including subsidiary rights, should be addressed to the Office of the Publisher, The World Bank, 1818 H Street, NW, Washington, DC 20433, USA; fax: 202-522-2422; e-mail: pubrights@worldbank.org.

Cover photo Chinese stock exchange image: © Wu Hong / epa / Corbis
Cover design Edelman Design Communications

ISBN: 978-0-8213-7672-0
eISBN: 978-0-8213-7673-7
DOI: 10.1596 / 978-0-8213-7672-0

Library of Congress Cataloging-in-Publication Data
   Innovation for development and the role of government : a perspective from the East Asia and Pacific region / edited by Qimiao Fan . . . [et al.].
       p. cm.
   Based on proceedings of the Innovation for Development Forum held in Shanghai, China, on Sept. 21–22, 2006.
   Includes bibliographical references and index.
   ISBN 978-0-8213-7672-0 — ISBN 978-0-8213-7673-7 (electronic)
   1. Economic development—Technological innovations. 2. Technological innovations—Government policy. 3. Finance—Technological innovations. 4. Competition—Government policy. 5. China—Economic policy. 6. Technological innovations—Government policy—China. I. Fan, Qimiao. II. Alternative Fuels Data Center (National Renewable Energy Laboratory) III. Title.
   HD82.I3417 2006
   338.9—dc22

                                                                                          2008037351

Contents

BOXES

FIGURES

TABLES

Foreword

Development is one of the major themes of today's world. In the context of global economic development practices, the development patterns of various countries fall primarily into three categories. The first is the resources-based pattern, which is supported by natural resource endowments. The second category is the dependency pattern, which is determined by a country's adjacency to economically developed countries with which it has close economic ties. The third is the innovation-based pattern, which is driven by innovation. Measured by levels of economic development, the current top 20 most developed countries in the world have opted primarily for an innovation-driven pattern. These countries share some commonalities: their overall innovation indexes are noticeably higher than other countries, their ratios of contributions made by scientific and technological progress are above 70 percent, their research and development (R&D) inputs as a percentage of GDP are above 2 percent, and their degrees of reliance upon external technologies are lower than 20 percent.

On a worldwide scale, there is a high positive correlation between innovation and economic development. Studies of economic development histories in various countries help to reveal some phenomenal growth trajectories. Two centuries ago, the U.S. economy was still very underdeveloped, but now its per capita GDP has almost reached the level of US$40,000. Another example is the Republic of Korea. At the beginning of the 1950s, Korea's GDP per capita was less than US$100, but by 2005 the figure had exceeded US$15,000. Praised as a Han River miracle, Korea is regarded as one of the Asian countries with relatively high innovation capabilities. A third example is China. As the world's largest developing country, China started to implement the policies of reform and of opening up in the late 1970s. China sustained a high economic growth rate and saw its per capita GDP increase from less than US$200 to over US$2,000 in less than 30 years, thus earning the name "China miracle" from the international community. China has also followed a development path characterized by sustained innovation that often started with pilot programs.

From a research perspective, there is no simple functional relationship between the innovation process and the factors that influence such a process. Despite the fact that the academic community has various definitions of *innovation*, in essence, innovation is a kind of evolution of ideas and processes with changes at the core. It aims to establish a "new function of production and achieve a new combination of production factors" in order to change the old economic development pattern that relies on the increase of traditional factors of production to a more sustainable one.

Such a process is undoubtedly a complex, dynamic one, with multiple factors influencing its progress, such as institutional arrangements, market structures, forms of industrial organization, R&D capabilities, intellectual property rights protection, and human resources development. Empirical studies show that factors influencing the innovation process vary noticeably from country to country, to a large extent owing to differences in areas such as social system, economic structure, history, and culture. On the one hand, such differences have increasingly enriched and deepened our research on the relationship between innovation and the factors influencing it; on the other hand, such differences inform us that development should be based on each country's respective social and economic conditions and characteristics and that each country should adopt an innovation pattern that is compatible with its economic development needs.

In a world with increased economic globalization, both developed and developing countries are faced with opportunities and challenges. If history can serve as a guide, experiences of economic development across the world have confirmed the critical role of innovation in economic and social development. Increasing a country's overall innovation capabilities and creating a sound and efficient national innovation regime help to increase the quality and sustainability of economic development and boost the country's core competitiveness. In the face of new challenges and opportunities, we need to join hands to think about, and further explore, how to develop relevant macroeconomic policies to help increase the innovation capabilities and to promote the healthy and harmonious development of the world economy.

Innovation for Development and the Role of Government can be viewed as a book resulting from the thoughts of some economic development trajectories in this region in relation to the issues mentioned above. This book is a collection and expansion of the speeches and remarks made by scholars, government officials, and international experts at the "Innovation for Development" forum held in Shanghai in 2006. From different perspectives, the book analyzes the close relationship between innovation and development stages as well as the increasing importance of innovation for sustainable development. In addition, from three perspectives, namely, the creation of an innovation regime, innovation-oriented fiscal and financial policies, and regional cooperation on innovation, and in two dimensions, namely, theory and practice, the book discusses and explores problems facing us all now and challenges in the future. The viewpoints in this book both reflect the research on the issues of innovation by its authors and, to a certain extent, mirror the views expressed by nonspeaker experts in the course of discussions at the 2006 forum.

Economic globalization is an inevitable trend. It is extremely necessary and valuable to conduct research on, and exchange views about, innovation and development against the backdrop of constantly deepening economic globalization. First and foremost, this helps us to see through the vast and complex economic surface to examine and discover the laws that drive sustainable economic development. Second, through sharing experiences of innovation among different countries, it can help us to establish a cooperative mechanism for innovation that can transcend social systems and cultural differences and promote the harmonious economic development for the region.

Yong Li
Vice Minister
China's Ministry of Finance

Preface

This book is a collaboration between the World Bank and the Asia-Pacific Finance and Development Center (AFDC) in China. It is based on the proceedings from the "Innovation for Development" forum, jointly sponsored by AFDC and the World Bank and held September 21–22, 2006, in Shanghai.

The main objectives of the forum were to (1) discuss the importance of innovation to development and poverty reduction, (2) explore fiscal and financial policies that are conducive to innovation, and (3) call for more regional cooperation in the protection and sharing of innovative knowledge. It was a high-level forum, and participants were senior government officials from economies in the Asia-Pacific Economic Cooperation, including a number of deputy ministers of finance and deputy central bank governors, policy researchers, private sector executives, and representatives of the Asian Development Bank.

The key issues discussed broadly included the relationship between innovation, competitiveness, and development; innovation and financial sector development; and innovation and government policies in China. Those three areas of investigation provide the overall structure for this book.

Acknowledgments

The editors—Kouqing Li and Runzhong Peng from the Asia-Pacific Finance and Development Center (AFDC), and Qimiao Fan, Douglas Zhihua Zeng, and Yang Dong from the World Bank—express their sincere thanks to the Chinese government, especially to Vice Minister Yong Li, Vice Minister Jun Wang, and Director-General Xiaosong Zheng from the Ministry of Finance of China, and to Dawei Xia, Director-General of AFDC; Frannie Leautier, former World Bank Vice President; and Alexander Fleming, Sector Manager of the Finance and Private Sector Development Division of the World Bank Institute, for their generous support of and guidance for the forum.

The editors thank all participants at the forum for their comments and discussions, which were very helpful in revising some of the chapters in this volume. They express their gratitude to Sheng Li and Tao Su, from the World Bank Beijing office, and to Hui Wang and Xiaohua Li, from the AFDC, for their excellent administrative support of the forum.

The editors are also grateful to Benjamin Kim Weng Chan for his assistance in editing three of the chapters that were originally written in Chinese, and to Stephen McGroarty, Janet Sasser, and John Didier in the World Bank for their publication assistance.

Contributors

Editors

Qimiao Fan	Lead Economist and Program Leader, World Bank Institute, World Bank, Washington, DC
Kouqing Li	Deputy Director-General, Asia-Pacific Finance and Development Center, Shanghai
Douglas Zhihua Zeng	Economist, World Bank Institute, World Bank, Washington, DC
Yang Dong	Consultant, East Asia and Pacific Region, World Bank, Washington, DC
Runzhong Peng	Deputy Director of Research Division, Asia-Pacific Finance and Development Center, Shanghai

Other Contributing Authors

Vivek Arora	Senior Resident Representative in China, International Monetary Fund
Chen Jia	Director General, Department of Enterprises, Ministry of Finance, Government of China, Beijing
Frannie Leautier	Former Vice President, World Bank Institute, World Bank, Washington, DC
Yong Li	Vice Minister, Ministry of Finance, Government of China, Beijing
Huaipeng Mu	Director General, Department of Financial Market, People's Bank of China, Beijing
Kaoru Nabeshima	Consultant, Development Economics Vice Presidency, World Bank, Washington, DC

Yaobin Shi	Director General, Department of Taxations, Ministry of Finance, Government of China, Beijing
Kenneth Waller	Senior Adviser, Asia-Pacific Economic Cooperation Business Advisory Council, Monash University, New South Wales, Australia
Jun Wang	Vice Minister, Ministry of Finance, Government of China, Beijing
Shuilin Wang	Senior Economist, Development Economics Vice Presidency, World Bank, Washington, DC
Shahid Yusuf	Economic Adviser, Development Economics Vice Presidency, World Bank, Washington, DC

Abbreviations

ABAC	APEC Business Advisory Council
AFDC	Asia-Pacific Finance and Development Center
APEC	Asia-Pacific Economic Cooperation
CBRC	China Banking Regulatory Commission
ESTD	early-stage technological development
FDI	foreign direct investment
GNI	gross national income
HTIZs	high-tech industrial zones
ICT	information and communication technology
IPO	initial public offering
IPR	intellectual property rights
IT	information technology
OECD	Organisation for Economic Co-operation and Development
OTC	over-the-counter (market)
PBC	People's Bank of China
R&D	research and development
S&T	science and technology
SME	small and medium enterprise
SOE	state-owned enterprise
VaR	value at risk
VAT	value added tax
VC	venture capital
WIPO	World Intellectual Property Organization (U.N. agency)

Introduction and Summary

Qimiao Fan and Douglas Zhihua Zeng

In today's highly globalized economy, innovation has become the key driver for growth and competitiveness. The capability to innovate and to bring innovation successfully to market will be a crucial determinant of the global competitiveness of nations over the coming decades. An innovation system consists of the network of institutions, rules, and procedures that influences the ways a country acquires, creates, disseminates, and uses knowledge. The actors in the innovation system include universities, public and private research centers, enterprises, consulting firms, policy think tanks, policy makers, and others. The innovation performance of a country largely depends on how these actors relate to each other as elements of a broader system.

The concept of *innovation* encompasses not only technological innovation, that is, diffusion of new products and services of a technological nature into the economy, but equally includes nontechnological forms of innovation, such as organizational or institutional innovations. The latter may include the introduction of new management or marketing strategies, adoption of new policies or creation of a new service (or nontechnological product), and improved approaches to internal and external communications and positioning. Innovation is important not only for the production sectors, where technologies play a crucial role, but also for the service sectors, such as finance and public services.

This book is the result of a joint forum on "Innovation for Development" held by the World Bank and the China-based Asia-Pacific Finance and Development Center (AFDC) in Shanghai in September 2006. The book examines the relationship between innovation, competitiveness, and economic growth; the role of innovation in financial sector development; and specific government policies for innovation in China.

The authors of book chapters are mostly policy practitioners and international experts on technology innovation or financial sector innovation. The book is intended to provide useful analysis on innovation and its determining role for growth and competitiveness and practical policy options for policy makers in the Asia-Pacific Economic Cooperation (APEC) countries/economies, especially those in China.

The opening address by Vice Minister Jun Wang emphasizes the importance of innovation to economic development using as an example China's rising demand for natural resources. He points out that the speed of China's economic growth has far outpaced this demand because China has constantly pushed ahead with technological and institutional innovations and reforms. However, given its vast population and scarcity of natural resources, China's future economic development will face far greater pressures. How to promote sustainable development through innovation is a huge challenge for the Chinese government. To cope with this challenge, the government of China is ready to introduce four major measures: (1) proactively offer guidance and support to promote indigenous innovation; (2) gradually establish a technology innovation system with enterprises as the cornerstone, the market as the guide, and industry, academia, and research institutions fully integrated; (3) provide more support to small and medium enterprises (SMEs); and (4) ensure better protection of intellectual property rights (IPR). In the end, he hopes that the Asia-Pacific region will share experiences with innovation, learn from one another, and make concerted efforts to promote this region's economic prosperity.

The keynote address by the former World Bank Vice President Frannie Leautier defines innovation as consisting of (1) new products, (2) new processes, (3) ways to penetrate new markets, (4) new supply sources or distribution methods, and (5) new industries. In addition to these five categories, the use of new management practices and organization structures, the development and retention of skilled personnel, and new ways of securing financial resources and managing interface with government and other external agencies are also forms of innovation. She acknowledges that innovation is a complex and multifaceted phenomenon and that there are many factors that influence a country's innovation performance; therefore, there is an important role for the public sector to play, particularly governments, in fostering innovation. However, innovation policies and ambitions have to be adapted to the levels of development and the institutional capacity of a country. She hopes that the forum will become a regular venue for South-South learning in the Asia-Pacific region.

The main contents of the book are divided into three parts. Part I, on innovation, competitiveness, and development, provides in one chapter a brief overview of the key determinants of innovation, the importance of innovation to competitiveness and development, and the role of government in fostering and promoting innovation. The key factors covered in chapter 1 include competition and market structure, IPR protection, quality and availability of human resources, investment in research and development (R&D), financing of innovation, technology diffusion systems, and the presence of industrial clusters. The chapter describes how governments can create an enabling environment for innovation by improving business climate, establishing well-balanced IPR systems, investing in human capital, enhancing the R&D infrastructure to attract private investment, encouraging the establishment of industrial clusters, and providing direct funding for basic research and developing a competitive financial market.

Part II of the book, which includes chapters 2 through 5, examines the relationship between innovation and financial development, and the importance of financial innovation for overall economic performance. Beyond generic analysis, this part also presents the cases of China and the APEC. In chapter 2, the first of two sections briefly synthesizes the contribution of key financial innovations and examines the role of information and communication technology (ICT) in accelerating the pace of innovation; the second section comments on the status of financial changes in China, the speed at which new financial instruments are being introduced into the Chinese financial system, and the areas where catch-up will have fruitful consequences. Four points that are covered in this chapter are worth noting. First, the financial system in the leading industrial countries is an unusually prolific source of innovations, and these are now spreading rapidly to industrializing countries. Second, as a result of the ICT revolution, the productivity of the financial sector has risen rapidly over the past decade. Third, financial innovations, which help to cut down transaction costs, promote liquidity, and better serve the requirements of borrowers and lenders, are making a significant difference to allocative efficiency and total factor productivity. Fourth, despite the great progress, China has yet to exploit the full potential of financial innovation.

Chapter 3 discusses China's experience with financial sector development and some of the challenges ahead in the context of international experiences. China has made significant progress in financial reforms in recent years; however, the financial system still favors large state-owned enterprises over smaller firms and private enterprises, and there is still much to be done to complete bank reforms and to further develop capital markets that can play a larger role in the financial system. A more developed financial system, with competition among banks, sound institutions, and developed capital markets, can help to allocate scarce resources efficiently, ensuring that they are used in the most productive way.

Chapter 4 provides an overview of product innovation and institution building in China's bond market over the past two years. The innovations of bond products in China include the securities companies' short-term financing bills, the subordinated bonds and ordinary financing bonds issued by commercial banks, credit asset securitization, the hybrid capital bonds, and the nonfinancial corporations' short-term financing bills. There are also great innovations in trading instruments and financial derivatives in the bond market. The major innovations in institution building include the establishment of the management mechanism for issuance of short-term financing bills, more stringent information disclosure requirements, and the introduction of the bond rating system. Driven by product innovation and institution building, China's bond market has been growing very rapidly in recent years; however, the development of corporate bonds still lags behind. Citing experiences of China's bond market in the past years, this chapter concludes that innovation is the driving force for the development of the bond market, which will contribute significantly to expanding the scale of corporate direct financing.

Chapter 5 discusses the key aspects of the finance sector that have been enabled by the introduction of innovation and technology, including efficiency, risk management, liquidity, pricing, regulation, and supervision. The advances in finance, in turn, have enabled and encouraged investment in technology and innovation in other sectors across an economy. This effect can be attributed to strengthened capabilities in risk management, financial modeling, product design, and proliferation of service forms. These key strengths promote technology and innovation, both at the firm level in real sectors and through their influence on specific markets for technology and innovation. This chapter also addresses the broader policy environment for financial innovation, such as strong legal, supervisory, and regulatory regimes, and the role regional cooperation can play in harnessing and facilitating innovation in the finance sector, especially in the Asia-Pacific region.

Part III, which consists of chapters 6 through 8, examines some specific government policies to promote innovation in China. Chapter 6 discusses the progress and challenges of China's enterprise innovation, and presents a set of policy recommendations. The author argues that the technology levels of China's industries and enterprises have been increasing, but compared with the advanced countries, the innovation capacity of Chinese firms is still weak. This can be attributed to the lack of efficient integration between the economic activities and science and technology, lack of effective policies and institutions that enable key industries to stimulate indigenous innovation, weak IPR protection, a performance benchmarking system that fails to emphasize indigenous innovation by firms, and more. To strengthen the innovation capacity among Chinese firms, the government needs to develop and implement innovation-oriented public policies and policy measures. Measures could include effective industrial policies to promote indigenous innovation, strengthen IPR protection, and support SME innovation through special programs. In addition, technology absorption, adaptation, and commercialization capabilities could be improved by establishing R&D platforms for industrial and core technologies, fostering science and technology talents, strengthening technology transfer mechanisms, and using direct public funds, tax-based fiscal incentives, and government procurement to promote indigenous innovation.

Chapter 7 explores the fiscal policy options for China. First, China's spending on R&D has been growing rapidly in past decades, but it's also important to raise the efficiency of R&D through organizational changes. Second, it might be more desirable to shift emphasis from direct budgetary supports of R&D to R&D tax credits, leaving the decision on what to do research on with firms. Third, experienced research managers in China are still scarce, so offering incentives to foreign companies to set up R&D facilities can be a potential means of training Chinese researchers; also, offering substantial incentives to Chinese researchers abroad is an effective way to attract back overseas science and technology (S&T) talents. Fourth, it is desirable to rigorously examine the gains from the multiplication of technology zones and the returns accruing from the fiscal incentives provided. Although the government incentives provided have led to a number of state-sponsored venture capital firms, the provision of capital remains subject to many administrative hurdles. Finally, because of the importance of high-quality S&T workers, fiscal support, at least for the leading universities, would be a sound investment.

Chapter 8 examines China's tax policy options for innovation. It looks at the challenges and opportunities faced by China regarding technological innovation and advocates that the government should adopt effective tax policies to promote technological innovation. On the basis of a brief assessment of China's current tax policy incentives, the chapter suggests some adjustments in the tax regime to encourage innovation. These include reforming the existing main tax categories and enhancing the catalyst role of tax policies in technology innovation. Greater efforts need to be made to transform the value added tax system from production based to consumption based and to allow technology firms to deduct fixed assets from taxable revenues. The existing preferential tax policies for innovation also need to be streamlined and improved to better promote fair competition, encourage the development of new and high-tech enterprises, strengthen the development of S&T talents, and spur the commercialization of S&T results.

Generally speaking, innovation covers a large spectrum of social, economic, scientific and technological aspects. The eight chapters included in this book reflect the different viewpoints on linkages between innovation and government policies and on the challenges that probably are commonly faced by economies in the APEC region, particularly for developing economies. Although this book is divided into three parts, it became clear in preparing the book that there is no single best way to sequence the chapters. The volume editors decided to follow a sequence that best illustrates the relationship between innovation and development, starting with the key conceptions about innovation, competitiveness, and development, followed by ideas on innovation and financial development, and finally consideration of the relevant government policies in China.

Two additional considerations are worthy of note. First, the reader will find some overlap among chapters. This reflects the authors' findings from different approaches to and perspectives on the same or similar topics. Second, this book should be viewed as a work in progress, and the opinions expressed are entirely those of the authors and do not necessarily represent the views of the World Bank or the AFDC.

Opening Address

Jun Wang

Distinguished guests,

I am very pleased to be here in Shanghai to attend the "Innovation for Development" forum organized by the Asia-Pacific Finance and Development Center (AFDC) and the World Bank, and to jointly explore and exchange views on the issues of innovation and development. First of all, on behalf of the Ministry of Finance of the People's Republic of China, I would like to offer a warm welcome to all the distinguished guests; sincere congratulations to AFDC and the World Bank, the forum's organizers; and heartfelt thanks to various economic bodies, international financial organizations, and friends that have supported AFDC for many years.

Innovation and development have always been important issues that the world has been paying attention to and actively discussing. Looking from a historical view of the world's economic development, one can see that innovation has always been closely linked with economic development, and each major leap forward in economic development has been accompanied by improvements in existing innovations and the emergence of new innovation. At a time when natural resources are becoming an increasingly salient constraint and bottleneck for economic development, countries are faced with the same problem of how to promote and facilitate the sustainable development of their economies through continuously increasing their national innovation capabilities.

From the end of the 1970s, China started to implement a reform and open-door policy, as a result of which its economy has registered a sustained and rapid growth. Not only has China's economic development improved the standard of living of the Chinese people and increased its overall national strength, it also has promoted the economic development of neighboring countries and regions and offered more investment opportunities and a vast marketplace for the rest of the countries and regions all over the world. It has thus played an essential role in promoting economic growth in the Asia-Pacific region and the world as a whole.

The high-speed growth of the Chinese economy has also attracted broad attention from the international community. Some countries have tried to establish a link between the high-speed growth of the Chinese economy and the rapid increase of natural resource consumption across the world. We can see that, along with the continued development of the Chinese economy, China's demand for natural resources has been on the rise; however, what's more noteworthy is that the speed of China's economic growth has far outpaced that of its demand for natural resources. Between 1990 and 2004, China's energy consumption grew on average less than 5 percent on a yearly basis, while the annual growth of the Chinese economy was as high as 9.3 percent on average. With a 5 percent growth of resources consumption, China has supported its nearly 10 percent rate of economic growth.

As the world's largest developing country, China still has high reliance on resources for its economic growth. The important reason that China could have supported its high rate of economic growth with a relatively low growth rate of resources consumption is that, while developing its economy, China has constantly pushed ahead with technological and institutional innovations and reforms. Since 1978, the Chinese government has introduced sustained reforms and innovation measures in the areas of institutions, technology, and administrative regimes. As a result of putting in place market mechanisms and promoting economic transformation, China has effectively improved the allocative efficiency of resources. The pilot program of administrative reforms in institutions of scientific research has greatly boosted enterprises' capabilities of technological innovation and enhanced overall industrial competitiveness. China's economic development practice has demonstrated that innovation is an important condition for the sustained and rapid growth of its economy.

Over the past two decades, the Chinese economy has made tremendous progress. China is, however, still a developing country with the biggest population in the world, with its per capita possession of natural resources far lower than that of the world average. In its future economic development, it is faced with far greater pressures in the form of resource constraints as compared with other countries. How to promote the sustainable development of its economy through innovation is a huge challenge facing the Chinese government over a considerably long period of time in the future.

China is in the process of implementing its 10th Five-Year Plan for National Economic and Social Development. It has advanced the concept of scientific development aimed to "shift China's economic growth from a resources-based pattern to an innovation-driven pattern and promote the notion of scientific development in both economic and social areas." In order to achieve these objectives, China will need to pay more attention to the driving role of innovation for economic development, put on top of its agenda the creation of a policy environment supporting and encouraging innovation, and adopt corresponding measures to boost the country's innovation capabilities. I would like to avail myself of this opportunity to brief our distinguished guests about the major measures that the Chinese government will introduce to encourage innovation.

First, the government will proactively offer its guidance and make great efforts to promote indigenous innovation. Efforts will be made to strengthen basic research, cutting-edge research, and technological research that offers social benefits, and to increase fiscal inputs for innovation in a number of major fields so as to achieve integrated innovation and leap forward in core fields and industries and to gradually form the country's innovation system. Toward this end, the government will foster a new innovation-supporting policy regime involving fiscal, taxation, financial, and human resources fields and create public goods that serve and promote innovation.

Second, the government will gradually establish a technology innovation system with enterprises as the cornerstone, the market as the guide, and industry, academia, and research institutions fully integrated. In recent years, through fiscal incentives and increase of fiscal inputs, China has encouraged the indigenous innovation by enterprises and made remarkable achievements. In the future, the Chinese government will regard support of the indigenous innovation by enterprises as an important policy objective of public finance, and it will further leverage fiscal policy measures such as increased fiscal inputs, taxation, treasury bonds, transfer payments, and fiscal subsidies, and establish a coordination mechanism for government procurement of self-innovated products, to support enterprise innovation.

Third, more support will be offered to small and medium enterprises (SMEs). SMEs have played an indispensable role for China's economic development, and they have also been an important driving force behind innovation activities. China will view the support of SME development as an important policy option for enhancing industrial competitiveness. The government will offer comprehensive fiscal policy support, including venture capital investment, tax incentives, and the creation of an SME credit guarantee regime and a public platform for technological services.

Fourth, more protection will be provided for intellectual property rights (IPR). China has successively promulgated, and accelerated the implementation of, a series of IPR laws and regulations, put in place a relatively complete IPR regime conforming to international norms, and conducted extensive international cooperation in the area of IPR protection. China will further strengthen IPR protection and offer comprehensive and complete legal protection for innovation.

Today, the Asia-Pacific region is a region with the greatest economic dynamism. Since the beginning of the new century, this region's economy has continued to show a strong development momentum. As a region relatively clustered with developing countries in the world, our economic development has, to a large degree, relied upon the continued increase of natural resource inputs. While we are sharing the success of economic development, we are also faced with challenges of resource constraints for our future development.

Economic development practice tells us that innovation is an inevitable choice for us to face the challenges. We are all actively exploring ways to promote economic growth through innovation. It is my hope that at this forum we can share our experiences on innovation, learn from one another, make concerted efforts to promote this region's economic prosperity, and together create a bright future for this region.

Thanks to you all!

Keynote Address

Frannie Leautier

Distinguished guests,

The issue of innovation for development has become a policy priority for many governments, not just in the Asia-Pacific region but over the world. Innovations can be broadly defined as new ideas or the process to undertake a change in one or more of many aspects of production, distribution, and consumption of economic goods. In more practical terms, innovation can be regarded as consisting of (1) new products, (2) new processes, (3) ways to penetrate new markets, (4) new supply sources or distribution methods, and (5) new industries. In addition to these five categories, the use of new management practices and organization structures, the development and retention of skilled personnel, and new ways of securing financial resources and managing interface with government and other external agencies are also forms of innovation.

Defined as such, innovation is a complex and multifaceted phenomenon. It is therefore not surprising that a large number of factors tend to influence a nation's innovation performance, and different factors may play differing roles in different countries. What is summarized here are some of the common factors that seem to influence innovation performance in general. The objective is not to provide a comprehensive exposition but to highlight a few issues for discussion:

1. *Competition and market structure.* A significant number of empirical studies show that, in general, a competitive market structure will help spur innovation. Competition will usually force companies to innovate (new and better products) to attract consumers and retain

market share while a monopolistic market structure will in general hamper innovation. The *key questions* here are, How can governments create a legal and regulatory framework that fosters the development of competitive markets? What is the role of the investment climate?

2. *Intellectual property rights (IPR) protection and incentives.* The protection of intellectual property rights is essential for stimulating innovation because inventors and firms will have no incentive to innovate if the fruits of their inventions are not protected. However, in some ways, too strict or too much protection of intellectual property rights may hamper innovation and the diffusion and reach of innovation, as in the case of pharmaceutical drug protection. The *vital issue* is to find the right balance between creating incentives for innovators on the one hand and encouraging competition and diffusion of innovation on the other.

3. *The availability and quality of human resources.* Quality of human capital, especially in science and technology, is of critical importance for innovation and a country's competitiveness. The issues here are not just the quantity and quality of basic and higher education but also the continuous upgrading of skills and knowledge. The *essential questions* are, How can countries ensure that they produce, attract, and maintain adequate supplies of well-trained scientists, engineers, and other professionals who will have not only the incentives but also the ability to innovate? What are the proper roles for the private sector and the public sector? Particularly important for developing countries is how they can attract and retain high-quality personnel in the presence of competition from developed countries? How can the knowledge and expertise of the diaspora be utilized? Here the Chinese experience of harnessing the role of the overseas Chinese community may be very instructive.

4. *Investment in research and development (R&D) and R&D infrastructure.* Innovation requires not only scientists and engineers but also a supporting infrastructure that provides the technological equipment, research inputs, financing, and other services. Empirical studies have shown that the amount and the quality of a country's investment in R&D and its associated infrastructure have an important impact on its innovation performance. In particular, R&D investment by firms will be critical. The *important questions* here are, What should be the type of R&D infrastructure and how much should a

country invest, given a country's level of development? How can such investment be financed? How much and what type should be financed by the private sector and how much by the public sector? What type of fiscal and financial policies should be put in place to support innovation? How can governments create a financial system that can provide multiple channels of competitive, market-based, and sustainable financing for innovation?

5. *Foreign direct investment (FDI) and industrial clusters.* FDI usually brings into a country new technologies and new products; and it has been shown to contribute to the gradual upgrading of a country's innovation capacity. Innovation is often born out of the blending of indigenous knowledge with the technological and organizational inputs from foreign firms. The recent development of China's electronic appliance industry, particularly the microwave, DVD, TV, and mobile phone sectors, is a good example of the role FDI has played in helping to move some Chinese firms from purely manufacturing businesses using foreign technology into innovative, global competitors.

The establishment of industrial clusters through market processes has also been shown to help improve a country's or region's innovation performance because of the presence of a critical mass of talents, the flow of ideas, linkages, and the availability of related services. The *key questions* here are, How can governments facilitate domestic firms to learn from foreign-invested firms, and how can the vast knowledge and innovative capacity of multinational corporations be harnessed to improve the innovation ability of local firms and be used to develop indigenous innovation? How can industrial clusters be fostered? And what type of infrastructure and other supporting services can the public sector provide to foster the development of industrial clusters that utilize a country's or a region's comparative advantages?

Clearly, there is an important role for the public sector to play, particularly governments, in fostering innovation. However, innovation policies and ambitions have to be adapted to the levels of development and the institutional capacity of a country. In many ways, China's and other countries' successes have shown that government support should probably be initially focused on the most promising regions and sectors where the country has comparative advantages in terms of the factors discussed above in order to build and demonstrate the confidence through success stories.

The South-South learning between developing countries is particularly important. We hope that this forum will provide an important opportunity for developing countries in Asia-Pacific Economic Cooperation to learn from each other and that the forum can become a regular venue for such learning. The World Bank Institute, the capacity-building arm of the World Bank Group, always stands ready to provide full-hearted support to facilitate such efforts.

PART I

*Innovation, Competitiveness,
and Development*

1

Innovation, Competitiveness, and Economic Development

Qimiao Fan, Yang Dong, and Douglas Zhihua Zeng

This chapter provides a brief synthesis of the key determinants of innovation, the importance of innovation to competitiveness and development, and the role of government in fostering and promoting innovation. Accordingly, the chapter is divided into three parts. The first part looks at key factors that determine a nation's innovation performance, including competition and market structure, intellectual property rights (IPR) protection, quality and availability of human resources, investment in research and development (R&D), financing of innovation, technology diffusion mechanisms, and the presence of industrial clusters. The second part discusses the importance of innovation to a nation's competitiveness and development. The last part examines the role of government in fostering innovation. This chapter is not intended to be comprehensive but to highlight key issues for consideration.

Determinants of Innovation

Many factors determine the level and effectiveness of innovation. Among these, the key factors may include the level of competition and market structure, institutional framework such as IPR protection, quality and availability of human capital, R&D investment, financing support such as the venture capital market, the technology diffusion system, and the presence of industrial clusters.

Innovation Defined

Simply defined, innovations are new ideas, or "the process to undertake a change in one or more of many aspects of production, distribution, and consumption of economic goods" (Beije 2000). In more practical terms, innovation can be regarded as consisting of (1) new products (material goods and intangible services), (2) new processes, (3) new ways to penetrate new markets, (4) new supply sources or distribution methods, and (5) new industries (Schumpeter 1912/1961, 66). In addition to these five categories, the use of new management practices and organization structures, the development and retention of skilled personnel, new ways of securing financial resources, and new ways of managing the interface with government and other external agencies are also forms of innovation (Mehta and Joshi 2002). Innovation is therefore the development, commercialization, and application of new and unproven technologies, untested products, and new services, processes, and institutions.

Determinants of National Innovation Performance

As defined above, innovation is a complex and multifaceted phenomenon, so it is therefore not surprising that a large number of factors tend to influence innovation performance. This section examines the common factors that determine a country's innovation performance, although different countries and regions have different local context. In particular, it looks at the role of the following factors: competition and market structure, intellectual property rights protections, the quality and availability of human resources, investment in research and development, venture capital support, technology diffusion, and industrial clusters.

Competition and Market Structure   In general, competition and a competitive market structure will encourage innovation because competition forces firms to continuously upgrade their products in order to retain and expand market share. Firms will maximize their expected profits under conditions of technological and market uncertainty; therefore, the incentives to innovate will depend to a large extent on the expected difference between post-innovation and pre-innovation rents of firms, namely, the incremental profits from innovation. This will be particularly true in sectors where firms are operating at similar technological levels. In the highly competitive sectors, pre-innovation rents are reduced by market competition. In sectors where innovations are made by laggard firms with already low initial profits, competition will mainly affect post-innovation rents, and therefore the Schumpeterian effect of competition will dominate (Aghion et al. 2002).[1]

However, some argue that competition will not always encourage innovation. A significant number of empirical studies by scholars suggest that there is a clear inverted U–shaped relationship between the two, as indicated in figure 1.1 (Aghion and Griffith 2005; Blundell et al. 2002).

As shown in figure 1.1, at the initial low level of competition, firms will innovate more as they face more competition, because if they do not and their competitors do, they will be worse off; at the higher level of competition, firms innovate less when the intensity of innovation increases (traditional Schumpeterian effect). The U.S. automobile industry, which is one of the most innovation-intensive industries, offers a good example of this effect (see box 1.1).

It should also be noted that competition-driven innovation can often result in duplication of efforts or in parallel research efforts. In cases where the innovator's efforts encounter enormous R&D costs and imitators can

Figure 1.1 *Inverted U–Shaped Relationship between Competition and Innovation*

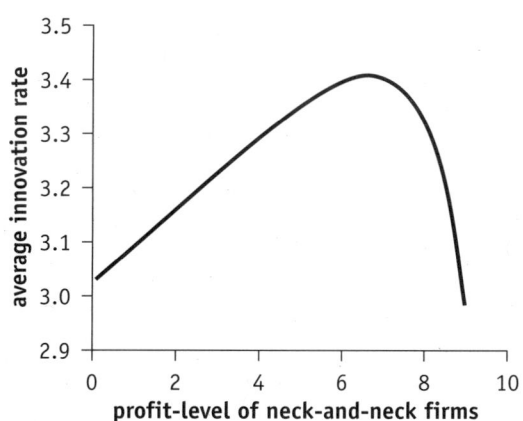

Source: Aghion et al. 2002.

Box 1.1 *The U.S. Automobile Industry: Competition Drives Innovation*

The U.S. automobile industry has undergone significant changes in the intensity of competition in the past decades. Since the 1980s, when the Japanese car companies started setting up assembly plants in the United States, the competitive environment in the U.S. automobile industry has changed remarkably, with far reaching consequences. Increased competition has reduced the industry's overall profit margins,[a] and market shares of the big-three U.S. car manufacturers (General Motors, Ford, and Chrysler) have decreased. Until 1980 the market was dominated by GM and Ford. During the 1980s, GM and Ford lost their market shares, mainly to Toyota but also to Honda. During the 1990s, the competition got very intense, and by the turn of the century GM, Ford, Daimler-Chrysler, and Toyota had roughly equal market shares, followed by Honda and Fiat. As a result, most of the firms in the U.S. automobile industry increased their R&D expenditures during the 1980s and most of the 1990s.

One might be tempted to see a simple positive relationship between competition and innovation in the industry; however, the evidence to support that idea is inconclusive. Both GM and Ford were increasing their R&D rapidly until the mid-1990s. Since then, GM has been spending less, and Ford's spending on R&D has roughly been flat. This trend continued until the mid-1990s. Since then, R&D investments by GM and Ford have declined. However, Honda and Toyota have persistently increased their R&D investment. Generally, for the big firms, market share and R&D are negatively related, and for the small firms they are positively related. R&D is generally highest for the firms with 20 percent to 30 percent of the market share.

The varied responses of firms to the changes in competition in the industry, in terms of their innovation activity, may be explained by the following. In a competitive environment, the market share of a firm will in large part depend on the relative quality and price of its products; therefore, the firm will invest in R&D to improve the quality of its products. The outcome of R&D is uncertain, but successful R&D leads to firms' increased technical knowledge. Increased knowledge translates into a better quality product. The investment in R&D could be regarded as a strategic decision: a firm takes the relative position of its rivals and their possible future actions into account before making its R&D decision.

Source: Hashmi and Biesebroeck 2006.

a. Profit margin is often used in the literature as a measurement of intensity of competition. Therefore, the decline in profit margins can be considered as an indication of an increase in competition.

easily take advantage of the original innovation, competition can also create a disincentive for innovation (U.S. FTC 2003).

Clearly, innovation will also depend on the market structure. Firms in a competitive market have more incentives to innovate in order to acquire or expand market share (Greenstein and Ramey 1998). Controlling for other factors, analysis shows that a monopolist that does not face the threat of entry has less incentive to engage in costly R&D to develop new products than does a firm encountering competition. Moreover, to the extent that the new products would cannibalize the monopolist's existing sales, the monopolist would be less likely to find R&D expenditures worthwhile (Arrow 1962). However, given the fear of losing profits, the monopolist that does face a threat of entry may have more incentive to invest in R&D than does a prospective entrant (U.S. FTC 1996).

Some empirical studies on the United States and other developed economies found a positive correlation between concentration or firm size and the measure of innovation. Economies of scale dictate that a large firm may benefit more from innovation, given the percentage cost reduction that applies to the larger volume of a large business, resulting in more benefits from innovation. Large firms will also be able to support a larger portfolio of R&D efforts, increasing the likelihood that they will develop an improved product or process and be able to market innovations more effectively.

Intellectual Property Rights (IPR) Protection Although the relationship between IPR protection and innovation is complex and not always linear, there is a growing consensus that IPR protection is essential for fostering more innovation and development. Scientific knowledge involved in innovation has some specific properties: uncertainty, inappropriability, and indivisibility (Lipsey and Carlaw 1998). In the absence of IPR protection, firms that develop new knowledge will not be able to appropriate the returns on their costly R&D investment. The differences between the social and private internal rate of return on R&D investment make it necessary for IPR protection, often in the form of patents, trademarks, and copyrights. There is evidence that strengthened IPR protections may stimulate innovation in developing countries as well. For example, in a survey of 37 Brazilian firms (conducted by the Brazilian Ministry of Industrial Development and Commerce and the American Chamber of Commerce), it was found that 80 percent of those firms would invest more in internal R&D and labor training if better legal protection were available (Sherwood 1990).

Patents are rights granted to inventors for a fixed period of time that allow inventors to exploit commercial revenues deriving from the application of their invention. Patents are one of the most widely used forms of IPR protection. Their use has a positive impact on market entry and firm creation, notably for small and medium-size firms that are usually less able to protect their innovations in alternative ways. Patents are also important to support investments devoted to the introduction of radical innovations, or those characterized by a high degree of uncertainty, high cost, and a long time lag between the invention stage and market introduction of the innovation (Crespi 2004). Furthermore, patents may enhance technology diffusion and aid in the commercialization of knowledge, which is also important for innovation.

In a series of surveys conducted in the United States, Europe, and Japan in the mid-1980s and 1990s, companies reported that patents are an important measure of protection that positively influences the amount of innovative effort in their industry, notably in biotechnology, pharmaceuticals, chemicals, and, to a certain extent, machinery and computers (Cohen, Nelson, and Walsh, 2000; Levin et al. 1987). Figure 1.2 suggests that countries that are regarded as high innovation performers also have

Figure 1.2 *Number of Patents Granted by the U.S. Patent and Trademark Office, 2006*

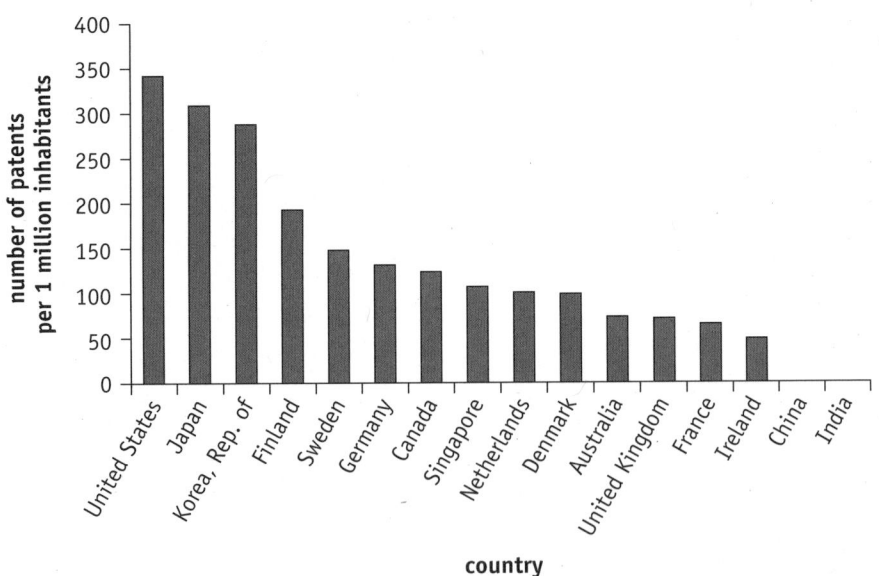

country

Sources: Based on data from World Bank (2007b) and the U.S. Patent and Trademark Office.
Note: In China and India, the number of patents granted by the U.S. Patent and Trademark Office in 2006 was about 0.74 and 0.46 per 1 million inhabitants, respectively.

higher patents per inhabitant rates in general. China's and India's numbers of patents granted by the U.S. Patent and Trademark Office are low in relation to population, but their absolute numbers are catching up rapidly. According to the World Bank's Knowledge Assessment Methodology (KAM), Switzerland, Sweden, Finland, the United States, and Denmark are currently ranked in the top five in terms of innovation (see table 1.1).

However, IPR protection may hamper competition, knowledge diffusion, and innovation if it is not structured properly or the protection period is too long. Patent protection may hamper further innovation when it limits access to essential knowledge, as may be the case in emerging technological areas when innovation has a marked cumulative character and patents protect foundational inventions. In this context, too broad a protection on basic innovation can discourage follow-on innovation if the holder of a patent for an essential technology refuses access to others under reasonable conditions (Bar-Shalom and Cook-Deegan 2002; Nuffield Council on Bioethics 2002; OECD 2003). Moreover, the cost generated by the IPR protection and the impact of potential market-manipulating behaviors to restrain competition may also cause some unjustified injury to the market and the consumer. Therefore, it is important to strike the right balance in structuring an IPR protection regime.

Quality and Availability of Human Resources   Human capital, especially in science and technology, is very important for innovation and knowledge-led economic growth. The long-term sustainability of the innovation process will depend on the adequate supply of well-trained scientists and engineers dedicated to the production of new technologies. Firm-level evidence indicates that the share of science and technology (S&T) workers in firms has an important impact on the introduction of new products and processes, irrespective of firm size and sector. S&T personnel help firms appropriate and apply the knowledge from basic research in industrial applications and raise the overall learning capacity of firms; therefore, they constitute a stock of intangible capital to the firm. Moreover, the movement of S&T personnel between sectors and firms and across national borders is very important for technology transfer. The skills developed by graduates in S&T provide substantial economic benefits to society as they move into the business sector. The knowledge they carry with them is especially important in newly emerging and fast-moving areas within the field of science and technology.

Finally, human resources in S&T are important because they constitute a source of entrepreneurs. Increasingly, academics and S&T graduates start

Table 1.1 *Global Comparison of Knowledge Economy Indexes Using Four Pillars: Top 15 Countries*

Country	KEI rank	KEI	EIR rank	Economic incentive regime	Innovation rank	Innovation	Education rank	Education	ICT rank	ICT
Sweden	1	9.26	13	8.59	2	9.72	6	8.98	1	9.76
Denmark	2	9.22	7	8.97	5	9.43	2	9.22	7	9.25
Norway	3	9.17	3	9.45	16	8.86	3	9.20	9	9.17
Finland	4	9.07	8	8.95	3	9.60	3	9.20	19	8.52
Netherland	5	9.02	11	8.69	6	9.41	8	8.74	5	9.25
Switzerland	6	8.99	4	9.42	1	9.82	34	7.44	4	9.28
Canada	7	8.94	5	9.38	8	9.35	10	8.62	23	8.40
Australia	8	8.88	18	8.39	19	8.71	5	9.17	6	9.25
United Kingdom	9	8.80	14	8.54	10	9.21	14	8.50	14	8.93
United States	10	8.80	16	8.45	4	9.44	16	8.35	13	8.95
New Zealand	11	8.76	12	8.60	20	8.69	1	9.30	22	8.44
Iceland	12	8.71	9	8.77	28	7.95	7	8.78	3	9.32
Austria	13	8.58	10	8.69	17	8.82	23	8.08	16	8.75
Ireland	14	8.56	15	8.54	15	8.92	11	8.62	27	8.16
Germany	15	8.54	19	8.38	13	8.93	22	8.08	15	8.79

Source: World Bank 2007a.

Note: The Knowledge Assessment Methodology (KAM) Knowledge Index (KI) measures a country's ability to generate, adopt, and diffuse knowledge. The index gives an indication of overall potential of knowledge development in a given country. Methodologically, the KI is the simple average of the normalized performance scores of a country or region on the key variables in three knowledge economy pillars—education and human resources, the innovation system, and information and communication technology (ICT).

The Knowledge Economy Index (KEI) takes into account whether the environment is conducive for knowledge to be used effectively for economic development. It is an aggregate index that represents the overall level of development of a country or region toward achieving the knowledge economy. The KEI is calculated based on the average of the normalized performance scores of a country or region on all four pillars related to the knowledge economy—economic incentive regime (EIR) and institutional regime, education and human resources, the innovation system, and ICT.

firms or participate in the creation of spin-offs from the public research sector, thereby contributing to knowledge and technology transfer and innovation (OECD 2000). According to World Bank research on knowledge's impact on long-term economic growth, an increase of 20 percent in a population's average years of schooling tends to increase the average annual economic growth by 0.15 percentage points (Chen and Dahlman 2004). Figure 1.3 shows that countries that are regarded as high innovation performers are the ones with higher rates of researchers per million people.

Investment in Research and Development Another factor that seems to differentiate countries' innovation performance is the investment in R&D by both the government and business sectors. All innovations will require R&D expenditures by both the government and business sectors. The question is, what should be the balance between government and business investment? On the one hand, it is ultimately the business sector that introduces and commercializes innovations (Porter, Furman, and Stern

Figure 1.3 *Number of Researchers in Research and Development, 2004*

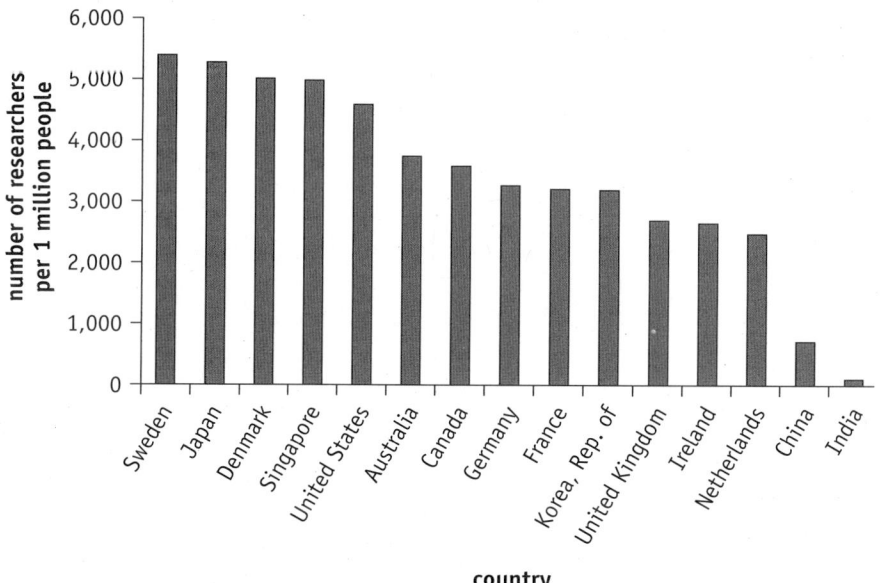

Source: UNESCO 2004.

2000). Therefore, efforts of governments and universities may support, but cannot be a substitute for, the technological efforts of the business sector. Studies also suggest that direct government support for industrial R&D has a smaller effect than private R&D expenditure (Griliches 1986). On the other hand, the business sector cannot generate by itself the optimal levels of R&D, thus depriving the economy of one of the key levers for sustained growth.

There are two main sources of market failures that inhibit the business sector from investing enough in R&D: (1) partial appropriability due to spillovers, which does not allow inventors to capture all the benefits of their innovation; and (2) information asymmetries, which lead to funding gaps (Goldberg et al. 2006). Figure 1.4 shows that, on average, the more innovative countries tend to have higher rates of R&D expenditure and a higher share of industry-related R&D spending (with industry-related spending ranging from 65 percent to 70 percent, and government spending from 20 percent to 30 percent [OECD 2002]).

Figure 1.4 *Countries' R&D Expenditures as a Percentage of GDP, 2004*

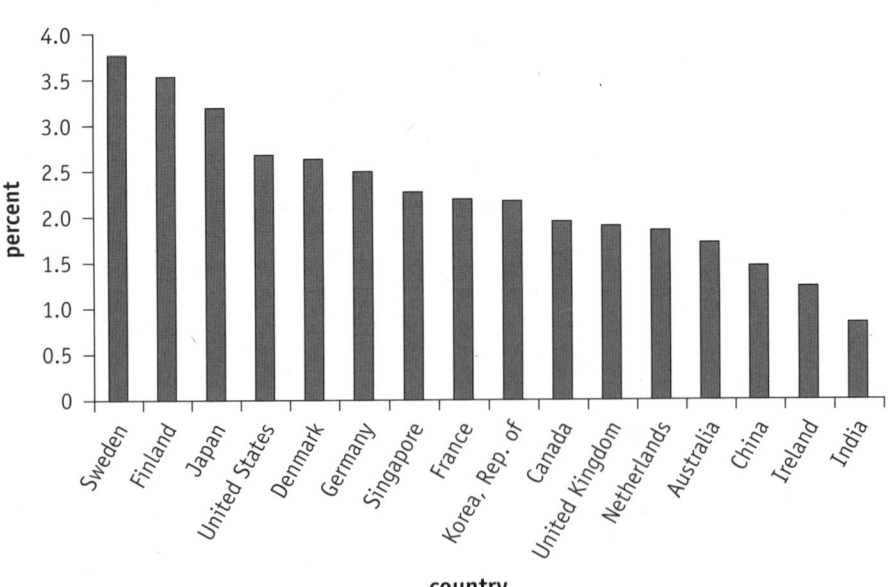

Source: UNESCO 2004.

As shown in figure 1.5, both internal financing by firms and government funding are very important in early-stage technological development (ESTD). One of the most notable features of the figure is that it demonstrates the absence of more mainstream intermediaries, such as banks, private equity, and other institutional investors in ESTD. For example, in the United States, one of the most innovative and prosperous economies, ESTD is undertaken directly, and financed by, firms or very specialized institutions, such as universities, with a significant role played by the government. Furthermore, internal funds account for the biggest share of ESTD financing in the United States, because that is the most straightforward way of addressing the information asymmetries.

Venture Capital Support   Venture capital (VC) plays an important role in the commercialization phase of the innovation chain. Venture capital targets projects that have passed the early stage; these projects may or may not have been supported by a grants program to reach the stage at which they are mature enough to be of interest to VC investors. On the other hand, it is important to note that, typically, purely commercial VC funds avoid the uncertainties connected with early-stage companies. To achieve

Figure 1.5 *Sources of Early-Stage Technological Development Funding in the United States*

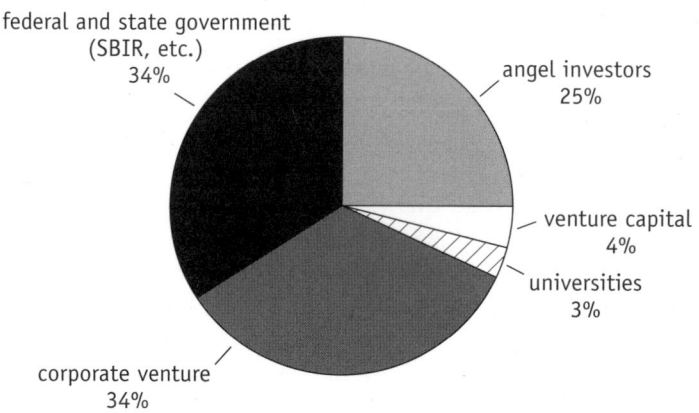

Source: Auerswald and Branscomb 2003.

Note: SBIR = Small Business Innovation Research program. "Angel investors" refers to successful entrepreneurs who look for new opportunities to invest private funds (earned from their own previous innovations) and are willing to invest in early-stage technological development projects in fields that they understand well.

the expected high commercial returns, the investors seek out companies that have successfully developed their innovation, proved its technical capability, and identified probable commercial applications and markets. At that stage, venture capital provides the funds to expand production and develop markets and the customer base and plays a critical role in supporting the later (and most visible) stages of commercialization.

Although venture capital plays a role in financing the commercialization of innovation and the expansion of innovative firms, it does not provide a solution to the market failure in ESTD. The success of VC funds depends on having a "deal flow" of attractive companies coming out of the ESTD phase. A venture capital program is therefore likely to work best in situations in which support for R&D through a grants program provides the critical funding at the earlier stages to advance companies to the level at which they can be supported by VC firms. The success of the most prominent VC funds in Organisation for Economic Co-operation and Development (OECD) countries relies, therefore, on three characteristics: investment expertise, risk profile, and deal flow (Goldberg et al. 2006).

VC investment analysts are highly specialized, with a strong understanding of different technology fields and their markets. If a VC fund invests in a company, it typically gains high levels of control and influence over the management decisions of the company. The VC fund manager brings management and commercialization expertise to the company and exercises control to ensure commercial success.

VC investment strategies are formulated so they can absorb a high number of failed investments. Typically, the VC fund aims to earn very high returns from one or two of 10 investments it makes, which compensates for the expected failure of the rest of the investments (cross-subsidization).

Venture capitalists rely on a supply of high-potential companies emerging from the earlier stages of business, technological, or innovation development. Therefore, venture capital works best in economies (such as those of the United States and Israel) in which the early stage of technological development is financed by internal funding, angel investors, or government-supported grant financing, or a combination of these.

In addition, the development of a VC market is closely linked with a well-functioning equity market. More than 24 countries operate separate boards and exchanges aimed at small and medium enterprises (SMEs). The key steps are to create an SME-friendly market architecture supported by effective institutions and to implement policies that foster a new class of investable equities. In the Republic of Korea, the venture promotion policy proved to be quite successful: in 2005, VC investments topped 0.25 percent of GDP, higher than the OECD average, and the number of VC-backed companies surpassed 10,000, which accounted for 5 percent of the country's total exports (Yoo 2007).

In China, the VC market is only at the initial stage. Government is the dominant source of capital, which compromises the incentive for fund managers to make high-risk investments, particularly in private enterprises. The legal framework for VC investment is also not complete. To promote the VC business, China needs to remove many regulatory barriers on VC firms and allow more exit options. In addition, more training needs to be provided to fund managers, and more transparency, objective performance measures, and external oversight need to be brought into the public VC firms (Zeng and Wang 2007).

Technology Diffusion For developing countries, the diffusion of technologies may be more important than the creation of new technologies. As the *Economist* (2007) observes, "India and China still have more to gain from the adoption and assimilation of technology than from invention per se. Some of their best minds are adding generously to the world's stock of knowledge, but the more urgent task for the countries themselves is to make wider use of know-how that already exists."

Technology diffusion involves the dissemination of technical information and know-how and the subsequent adoption of new technologies and techniques by users (Tassey 1992). In this context, technology includes both "hard" technologies (such as computer-aided machine tools) and "soft" technologies (for example, improved manufacturing, quality, or training methods). Diffused technologies can be embodied in products and processes (Shapira and Rosenfeld 1996). Although classic models may suggest a straightforward, linear path from technology creation to commercialization and adoption, in practice, technology diffusion is more often a complex and iterative process (Edquist and Jacobsson 1988; Maleki 1991).

There are various technology diffusion programs, such as awareness building and technology demonstration, technical assistance and consultancy, collaborative research and technology projects, facilities or agents for technology transfer, interfirm cooperation, standardization, procurement, and technological information services. In general, there is a great role for government to play in supporting the dissemination of knowledge and technology; however, the government should not directly support or push the adoption of technology by enterprises or other economic agents. Instead, it should focus on creating an environment that is conducive to facilitating development, commercialization, and use of improved technologies, including establishing appropriate institutions. Some good practices of technology diffusion programs are summarized in box 1.2.

In the case of China, the technology diffusion system is still quite weak. To effectively disseminate the R&D results from elite labs or universities

---

Box 1.2 *Best Practices for Efficient Technology Diffusion Programs*

The experience accumulated in both industrial and developing countries has highlighted four important general principles of technology diffusion:

1. The status of disseminators should be clearly recognized wherever they operate (in specialized centers, universities, and so forth). They should be adequately remunerated for the services they provide and not penalized in their careers (for example, academics involved in technology extension). In addition, the private sector's involvement in technology consulting services should be encouraged.

2. The appropriate status of a dissemination entity is generally a nonprofit organization. In China, such a status has been adopted by certain efficient organizations in, for instance, head offices of science parks, but this status needs to be strengthened from a legal perspective.

3. A minimum guaranteed core funding is necessary for these organizations to operate efficiently in the long term. The amount will vary with the specific public good and the level of privatization. In most cases, core funding should start with at least 30–50 percent of the budget, and higher when the organizations have to operate in depleted or developing areas or are involved in research. As demand takes shape, the proportion of core funding from the sponsoring government institutions can be reduced gradually.

4. Operations of a significant size generally should be financed jointly by the central government and the local or provincial authorities, with the funding of the former conditional on the capability of the latter to respond with a similar amount. Sometimes, alternatively or in addition, private business resources should join the programs to fund infrastructure elements, such as buildings, equipment, and personnel.

Source: Adapted from World Bank 2001.

or from technology parks to the production sector, China needs to put massive effort into promoting or strengthening technology and business incubators and regional clusters, engineering research and productivity centers, agricultural and industrial extension services at various levels or regions, and technical norms and standards (Zeng and Wang 2007).

Industrial Clusters Innovation is also determined by the microeconomic environment present in a nation's industrial clusters. Industrial clusters include four key elements: high-quality and specialized inputs, a context that encourages investment and intense rivalry, pressure and insight gleaned from sophisticated local demand, and the presence of cluster-related and supporting industries (Porter, Furman, and Stern 2000). As a result of an interactive learning process, innovation is deeply embedded in the relationships between the firm and its environment. Therefore, firms located in clusters can "plug into" the localized knowledge externalities, specialized labor markets, and the dedicated institutional support system and use these resources for maintaining an innovation-based competitive advantage. More specifically, industrial clusters can facilitate innovation through knowledge and technology diffusions through interfirm linkages, skills mobility, joint research and collaborations, as well as risk sharing (Zeng 2008). Clusters can also spur innovation by reducing search costs in general, especially the search costs of employers for workers and of workers for the jobs. This will therefore make employers more willing to invest in technology-specific human capital and increase the chances of innovation (Morck and Yeung 2001). The Silicon Valley, a world center of computer, software, and Internet industries, offers a good example of this effect (see box 1.3).

Pairing of certain clusters may also benefit innovation as a result of knowledge spillovers and other interrelationships. Just as a strong cluster innovation environment can amplify the strengths of the common innovation infrastructure, a weak one can stifle it. For example, despite strong infrastructure supporting scientific education and technical training in France, national regulatory policies toward pharmaceuticals limited innovation in the French pharmaceutical cluster during the 1970s and 1980s (Thomas 1994).

The relationship between the common innovation infrastructure factors described above and industrial clusters is reciprocal. For a given cluster innovation environment, innovative performance will tend to increase

---

Box 1.3 *Establishment of Industrial Clusters: Silicon Valley*

An industrial cluster refers to a geographic concentration of companies, colleges, and research labs aiming to achieve synergy in terms of sharing the results of technology development, human resources, and information. Innovation and entrepreneurship can be supported by a number of mechanisms operating within a cluster, such as easy access to capital, knowledge about technology and markets, and collaborators.

Silicon Valley is a world center of the computer, software, and Internet industries. It started in the 1950s with a modest plan by Frederick Terman, the dean of Stanford's Engineering School, to create an industrial park on unused Stanford land. The region took off in the 1970s with the development of the personal computer by Apple Computer Inc. and others, and it has exploded since then with the creation of the Internet and the enormous demand for software. Silicon Valley companies now employ more than 1 million people, almost 40 percent of whom have at least a bachelor's degree, and more than a third are foreign born. They are attracted by the good jobs and by the early access to frontier developments in high-tech fields.

Faculty and graduates of the strong science and engineering departments of two nearby universities, Stanford and the University of California at Berkeley, have been leaders in forming dynamic start-ups. A large pool of engineers, scientists, and software experts are available to both new and old companies. Information about new technical and market opportunities flows through institutions and informal networks very rapidly. At Silicon Valley, entrepreneurs find that access to capital is easier in the cluster, and venture capitalists and investment bankers find it easier to locate new investment opportunities. Firms in Silicon Valley participate in thick markets for technical labor, managers, capital, and other inputs. Many of these benefits arise by capturing external economies, which lowers the costs of invention and growth at large scale.

Source: Bresnahan and Gambardella 2004.

---

with the strength of the common innovation infrastructure and vice versa. The quality of linkages influences the extent to which the innovation potential induced by a common innovation infrastructure is translated into specific innovative outputs in a nation's industrial clusters, thus shaping the realized rate of national R&D productivity. Linkages can be facilitated by various types of institutions, ranging from universities to cluster trade associations to informal alumni networks (Porter, Furman, and Stern 2002).

Innovation, Competitiveness, and Economic Development

In today's highly globalized world economy, innovation has become the key driver for competitiveness and economic development. Only through innovation can enterprises enhance their productivity and meet the increasing and constantly changing market demand, and countries thereby achieve sustainable economic growth. Here, the concept of competitiveness and the relationship of innovation with competitiveness and economic development are briefly discussed.

Competitiveness Defined

Competitiveness in this context means possessing the capabilities needed for sustained economic growth in an internationally competitive environment in which there are other countries, clusters, or firms that have an equivalent but differentiated set of capabilities of their own. *Competitiveness* also implies a continuing rise in the living standards of individuals that are members of a social group with the required capabilities (Tyson 1992). Moreover, *competitiveness* can be regarded as entailing the comparison of relative growth rates or benchmarking of performance to assess how well each participant has done in developing the capacities for innovation and growth, rather than being about the mutual potential for damaging one another (Krugman 1994, 1996).

Linkages between Innovation and Competitiveness

Competitiveness comes from the creation of the locally differentiated capabilities needed to sustain growth in an internationally competitive environment. Such capabilities are created through innovation. It is important to note that competitiveness is relative, and maintaining competitiveness means different things in different contexts. For firms in high-income countries, being competitive may require having a significantly more attractive product or a better production process than is needed by firms in low-income countries. For the latter, being competitive may not require being at the frontier, but rather learning foreign technology and its diffusion, and adapting it to local circumstances to meet demand. However, for all countries, continuing innovation is an imperative for staying competitive if technological advances in the industry are significant (Nelson 1993).

Table 1.2 Ranking of Innovation and Competitiveness, 2006–07

Economy	KAM Index on Innovation	Global Competitiveness Index	Economy	KAM Index on Innovation	Global Competitiveness Index
Switzerland	9.82	2	Japan	9.17	8
Sweden	9.72	4	Germany	8.93	5
Finland	9.60	6	Ireland	8.92	22
United States	9.44	1	Australia	8.71	19
Denmark	9.43	3	France	8.64	18
Netherlands	9.41	10	Korea, Rep. of	8.44	11
Singapore	9.40	7	Russian Federation	6.92	58
Canada	9.35	13	Thailand	5.95	28
Israel	9.32	17	China	5.09	34
United Kingdom	9.21	9	India	3.93	48

Source: Authors' compilation from World Bank 2007a and the World Economic Forum 2008.
Note: A 10 is the highest score for the KAM index.

Though there is no uniform theory that provides a simple picture of the linkages between innovation and competitiveness, what is clear is that the world's most innovative economies are also among the top performers in terms of global competitiveness. Table 1.2 shows the World Bank's Knowledge Economy Index and the World Economic Forum's Global Competitiveness Index for 20 developed economies. In general, the more innovative an economy is, the more competitive it is.

Innovation and Economic Development

Innovation, notably technological innovation, is the major driving force in economic development. It contributes to economic development through enhanced effectiveness of physical capital; complementarities with labor; magnified productivity of human capital; and impetus for further investment in intellectual, physical, and human capital (Willig 2006). Moreover, innovation spurs growth through the diffusion of technology from the developed to less-developed countries. Theoretical and empirical evidence demonstrates the positive correlation between innovation performance and economic development. Recent studies indicate that technological progress is the cause of more than one-half of the growth of the U.S. economy (Schacht 2000). Because of factors such as globalization, increasing competition, the growing impact of information and communication technology, and the fast pace of scientific and technological change, firms must innovate more rapidly than ever before. As shown in figure 1.6, an economy's innovation ranking and level of economic development, measured by gross national income (GNI) per capita, are related. China, however, is an exception. Compared to other economies with a similar GNI per capita level, China's innovation performance is much higher than average.

Government's Role in Innovation

Given the preceding discussion on the determinants of innovation performance, it is clear that government can play an important role in shaping a country's innovation performance. Governments can create an enabling infrastructure for innovation by improving the business environment, establishing well-balanced IPR protection systems, investing in human resources, enhancing the R&D infrastructure to attract private investment, and encouraging the establishment of industrial clusters. Governments

Figure 1.6 *Correlation between Innovation and Income Level, 2007*

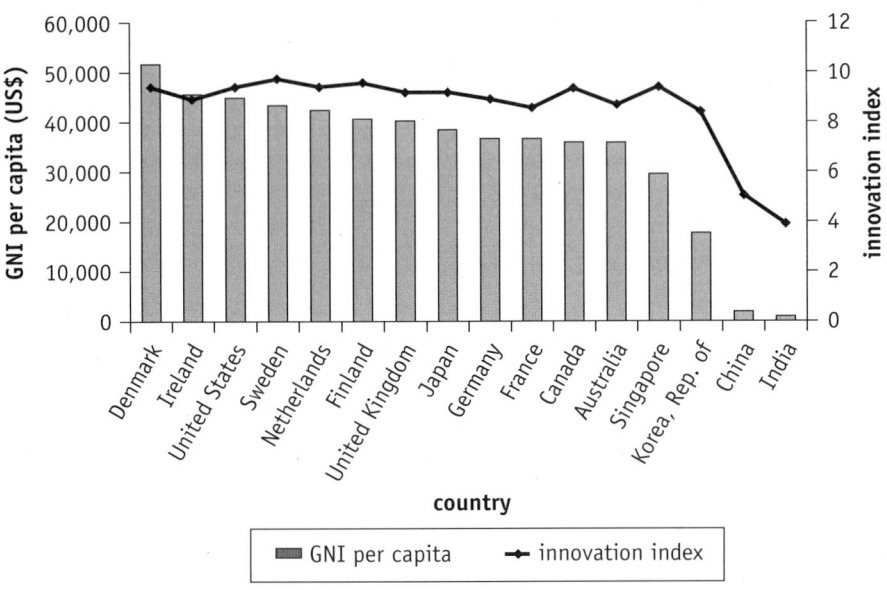

Source: World Bank 2007a, 2007b.

can also help increase the financial resources available for innovation by providing direct funding for basic research and developing a competitive financial market. Countries that improved their innovation performance over the past decades—Finland, Germany, Korea, Singapore, and the United States, to name just a few—have all implemented sound public policies in shaping a greater national innovation capacity. Box 1.4 illustrates how the government of Korea played a critical role in promoting the country's innovation capacity.

***

Box 1.4 *The Republic of Korea: The Role of Government in a National Innovation System*

Since the 1960s, Korea has transformed its industrial structure and caught up with developed countries. Its achievements over the past four decades were hailed as successful in terms of economic development. Korea is regarded as a model of successful technology catch-up. It began as an imitator in the 1960s and became a leading innovator in several high-technology areas within a generation. The experience of Korea confirms that catch-up through imitation requires conscious efforts and well-designed policies. The

Box 1.4 *(continued)*

government took a very active role in upgrading industry's technological capabilities. The Korean National Innovation System (NIS) has evolved in accordance with the development of the economy and industry and the accumulation of technological capabilities. Its evolution can be described in terms of the five stages shown in the table.

Periods	Characteristics of Korean innovation policies
1960s (infant stage of S&T policy)	Beginning of scientific education
	Beginning of science and technology infrastructure construction
1970s (institution building)	Construction of government-sponsored research institutes
	Technical, scientific, and further education
	Beginning of industrial R&D
1980s (National R&D program)	Promotion of key technologies through national R&D program
	Activation of industrial R&D
	Mass production of highly qualified R&D personnel
	Expansion of S&T-related ministries
1990s (diversification of government R&D)	Expansion of R&D resources and their efficient utilization
	Promotion of academic innovation potentials
	Introduction of regional innovation policy
	Introduction of Research Council system
2000s (elaboration of government R&D)	Enactment of Basic Law of Science and Technology
	Selection and concentration on major technologies
	Coordination of innovation policies
	Basic research and welfare technologies

Since 2004 the Korean NIS has undergone a revolutionary change owing to factors such as the need to reinforce the transition to an innovation-driven economy, the government's vision of a science- and technology-based economy, and the need to improve the effectiveness of the NIS. The reform is based on the concept of the third-generation innovation model, which emphasizes integration and coherence of science, technology, and innovation policy.

Source: Hong 2005.

Balanced IPR Protection

Intellectual property rights protection plays an important role in spurring innovation by creating the incentives for investments that are needed to develop new innovations and by ensuring that the inventors in innovation can reap the benefits of innovation. As discussed earlier, however, IPR protection that is too broad and too long may in some cases inhibit the diffusion of knowledge and innovation. In a consideration of IPR protection, there are trade-offs between incentives for innovation, on the one hand, and competition in the market and diffusion of technology on the other. Both the positive and negative impacts of IPR protection need to be carefully considered in developing and adopting IPR legislation and regulations.

Creating a balanced IPR regime is necessary but not sufficient to maximize the competitive gains from innovation. The IPR policies must be embedded in a broad set of complementary policies that optimize the effectiveness of IPRs and foster further innovation and development (Maskus 2000). Among such policies are developing human capital through education in science, technology, management, and law; encouraging the acquisition of skills; developing competitive financial systems, promoting research-industrial linkages; and establishing industrial clusters. Singapore, one of the top performers in growth and competitiveness in 2005, provides a good example (see box 1.5).

Investment in Human Resources and R&D Infrastructures

Countries that have a large enough pool of skilled, flexible, mobile, and networked individuals have a competitive edge in the global market. As shown in figure 1.7, the world's top performers in innovation all have a high ratio of R&D expenditure to GDP. In addition, these countries are also successful in converting R&D investments and educational capacity into industrial competitiveness. Many OECD countries have recently implemented policies to strengthen the quality and availability of their human resources (see box 1.6). The European Union is undertaking concerted efforts to increase investment in R&D activities to 3 percent of GDP by 2010, to be in line with the Lisbon goals and the creation of a European Research Area. In Asia, the Korean government also offers a good example of investing or encouraging investment in R&D in order to remain at the forefront of innovation (see box 1.7).

Box 1.5 *Creating an IP Culture in Singapore: Using Proactive Policies*

Singapore recognizes the importance of intellectual property (IP) to its economy, both as a national resource and in efforts to attract foreign investment. To develop intellectual property as a strategic and competitive asset, Singapore adopted a proactive IPR policy for the development of high-value-added and creative-content industries. In 2000, the Intellectual Property Office of Singapore (IPOS) was converted into a semiautonomous statutory board that was charged with administering the IPR system in Singapore, among other things. One of the recent IPOS initiatives is the provision of intellectual property information via the recently launched IP portal, SurfIP (http://www.surfip.gov.sg). The portal allows searches across multiple patent databases in various jurisdictions, as well as provides other technical and business resources. On the IPR enforcement front, the agency primarily responsible for domestic enforcement is the Intellectual Property Rights Branch, a specialized crime division of the Criminal Investigation Department, and border enforcement is undertaken by the Customs and Excise Department. In the field of education, Singapore has public education campaigns led by IPOS and the National Science & Technology Board and aimed at promoting greater public awareness of IPR. Today, Singapore is one of the leading nations in terms of patent filings and the creation of other IP assets.

Source: WIPO 2003.

Figure 1.7 *R&D Expenditure as a Share of GDP, 1980–2004*

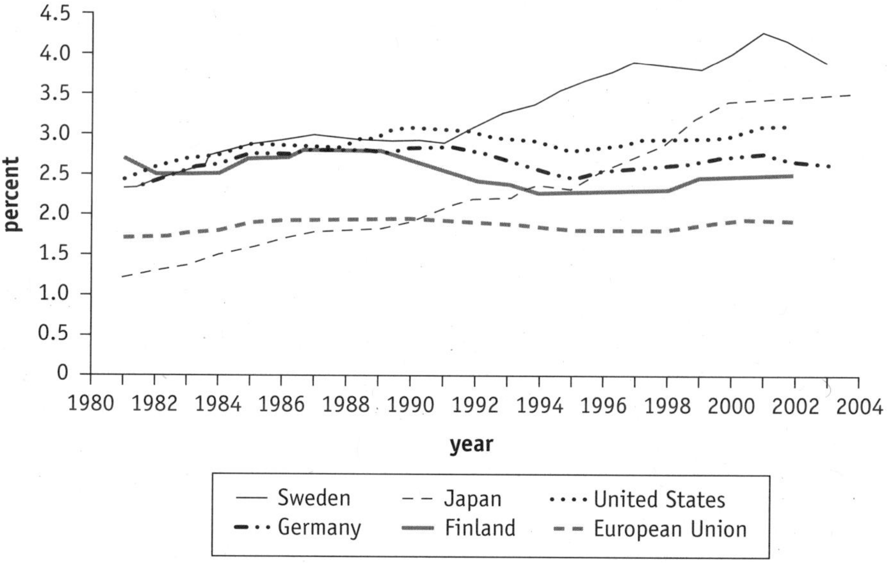

Legend:
— Sweden   – – Japan   •••• United States
—•• Germany   — Finland   – – European Union

Source: OECD 2005.

***Box 1.6** Investing in Human Resources and R&D: Recent Policy Development in Some OECD Countries*

Norway. The Ministry of Education, Research and Church Affairs has announced policies to improve recruitment in the public sector. The ministry is creating an average of 150 posts per year over five years, notably in priority areas such as medicine, informatics, and law. The ministry also supports measures to increase the numbers of women in natural sciences and technology, and to increase their share in permanent posts, particularly at the professorial level. Finally, a new program for workplace training has also been launched.

Finland. The government has launched a public-private partnership program whereby industry contributes to the training of new information technology (IT) graduates by putting equipment and experts at the disposal of educational institutions, offering internships, and encouraging their interns to graduate. The program involved over 20,000 students between 1998 and 2002 and is expected to increase the number of degrees by one-third between 1999 and 2006. To increase the pool of potential students, an additional program is being launched to strengthen education in mathematics and science. Measures are also being taken to find ways to attract more female students to the fields and alleviate the shortage of qualified teachers.

United Kingdom. The Faraday Programme promotes a continuous flow of industrial technology and skilled people between industry, the universities, and intermediate research institutes. A key element of the concept is the training and development of doctoral students who combine technical expertise with commercial awareness by undertaking industrially relevant research within intermediate institutes. In 1999, the Faraday Programme was to be expanded, with a focus on entrepreneurial activities and research commercialization.

Source: OECD 2000.

Fiscal and Financial Policies That Support Market-Led Innovation

The generation of new knowledge generally entails high costs. If R&D expenditure is unlikely to result in higher profits for the firm, there will be a strong disincentive to invest in innovation in the first place (Burrone and Jaiya 2004). Particularly in the initial stage of an innovation, the eventual success of innovation is highly uncertain, and potential investors are risk-averse with regard to financing its development. As seen earlier, the government can play an important role in such cases.

Box 1.7 *The Republic of Korea: Promoting R&D and Innovation*

It was only in the early 1980s, stimulated by changes in economic environments, that Korea embarked on serious efforts to develop indigenous R&D. Industrial development had reached such a stage that Korean industries could no longer rely on imported technologies and cheap domestic labor to compete in international markets. To meet the challenge, Korea required a supply of highly trained scientists and engineers, as well as financial resources, to support R&D activities that are by nature uncertain and risky. Korea was fortunate in this respect, because, thanks to Koreans' aspiration for education, the country had a large pool of scientists and engineers and also because the large Korean conglomerates were financially able to venture into new technologies. By the early 1990s, Korea's R&D investment exceeded 2 percent of GDP, of which private industries accounted for the bulk. The government had encouraged the private R&D investments through an outward-looking development strategy (export driven) and favorable policies towards large firms called the *chaebol*, which enjoy greater financial affluence.

The Asian financial crisis of 1997 changed the whole dynamism. Private businesses responded to the crisis by severely cutting R&D investments. To counteract the economic effects of declining R&D investments in the private sector, the government increased R&D spending to 5 percent of its budget, focusing on the development of IT and related industries. During this period, IT sectors played key roles in innovation in Korea, leading the country's recovery from the economic crisis as well as its move toward a knowledge-based economy.

Source: Chung and Shu 2007.

In most countries, governments have a major responsibility for the funding of basic research, although countries differ significantly regarding how much of that type of funding is needed, where most of the basic research would be carried out, and the kinds of applied research and development the governments might finance (Nelson 1993).

In general, government supports have taken two forms. The first approach is direct government support for firms' R&D investment, typically at the early stage of technological development (see box 1.8). This can be through procurement preferences, tax incentives, and direct grants or loans. The second approach is intervention in the market, through direct government funding or incentives for private venture capital funding, for financing innovative technology–oriented firms that are engaging in commercialization of R&D. In the most innovative and developed economies,

Box 1.8 *Fiscal and Financial Policies: The Case of TEKES in Finland*

In 2004 most of the support funds (82 percent) administered by the National Technology Agency of Finland (TEKES) were in the form of direct grants. By 2004 TEKES provided 42 percent of all of its funding through technology programs, totaling 171 million euros (about US$266 million). In total, TEKES has been funding 26 technology programs, focusing on a broad variety of technology sectors that range from public health care technologies to nanotechnology and business and management technology. The economic rationale of the technology programs is to enhance R&D cooperation between different companies, public R&D institutes, and international actors and to transfer knowledge and skills among the participating entities. Internal evaluations of initial programs find positive returns to the promotion of R&D cooperation and coordination. However, the success of the programs relies equally heavily on the quality of public administration in identifying and deciding on relevant program areas. The decisions to launch specific technology programs are based on strategic decisions within TEKES.

In interviews conducted by World Bank staff, TEKES decision makers emphasized that TEKES relies heavily on information feedback mechanisms and coordination with local R&D institutions and industry in formulating the technology strategy. Although the process is not formalized, the cooperative model of public policy formulation in Finland fosters a bottom-up approach to developing the technology program priorities, thereby avoiding some of the risks of top-down industrial policies.

TEKES represents international best practice in the support of innovation through grants and other soft funding instruments. The diversified approach to different project stages, the maturity of companies, and the emphasis on grant funding for start-ups and projects with high technological and commercial risks are of direct relevance to many countries. However, the decision-making process, project selection, and formulation of programmatic priorities are heavily dependent on the quality and capacity of public servants, as well as on the virtual absence of corruption and rent seeking and a transparent and cooperative approach to public policy formulation. Given these requirements, the TEKES model is not likely to be transferred as is to many developing countries. Governments should focus on adopting the funding instruments but complement the decision-making processes with independent, external control and oversight through peer reviews, foreign experts, and so forth. In regard to programmatic priorities, governments should be encouraged to emphasize neutrality in the early stages of the development of funding programs and to focus on technology policies based only on ex post patterns occurring over time.

Source: Armas et al. 2006.

such as Finland and the United States, policy packages of government support and intervention are widely credited with helping to support the technology-oriented firms to maintain high levels of innovation and growth (Goldberg et al. 2006).

Strengthening of Research-Industry Linkages

To build their own innovative capability, firms need to access external sources of information, knowledge, know-how, and technologies. Collaboration with universities and S&T institutes is a primary way for firms to tap into knowledge and technologies. There are many ways that government can help strengthen the research-industry linkages. One is to strengthen the technology "brokering" programs. Innovation brokers, if properly operated, can play very instrumental roles in stimulating less innovative firms to start a learning process to become more innovative, as well as in promoting the formation of networks and interactive learning among firms and knowledge institutions (Zeng 2007).

A relevant example is the Norwegian program for Technology Transfer from Research Institutes to SMEs (TEFT), which uses technology attachés as brokers (see box 1.9). In performing their roles as analysts of firms, brokers, and mediators, these attachés can be seen as prototypes for a proactive working method. The program is able to mobilize less innovative SMEs to cooperate more with knowledge institutions through networks and to perform joint innovation projects. However, it should be noted that the formation of such networks is a long-term process that requires active development of mutual interests and trust from both parties (OECD 2004). Other means include encouraging universities and R&D institutes to conduct more joint R&D activities with enterprises, and developing personnel mobility schemes, among other projects.

In addition, enhancing the innovative and absorptive capacity of firms should be equally important. Firms need to acquire new knowledge and technology, either domestic or foreign, and to absorb it on a continuing basis. From the innovation systems perspective, this means improving the firm's ability to access the appropriate networks, to find and identify relevant technologies and information, and to adapt such knowledge to their own needs. Government should focus not just on upgrading the abilities of individual firms but also on enhancing the networking and innovative performance of clusters of firms and sectors (OECD 1997).

Box 1.9 *The TEFT Technology: Attachés as Brokers*

Started in 1994, TEFT (Technology Transfer from Research Institutes to SMEs) is a nationwide program run by the Research Council of Norway (NFR). The NFR encourages SMEs to become more R&D conscious by developing closer links between such companies and the national technological R&D institutes. TEFT targets two groups of institutions. The first group is the non-R&D-intensive SMEs with weak internal resources; the second group includes the five largest polytechnic R&D institutes in Norway. The aim of TEFT is to change the R&D institutes' attitudes toward SMEs and strengthen their knowledge about SMEs' innovation needs.

The program spans the whole spectrum of Norwegian industry, but it is primarily intended to reach sectors with low or average levels of R&D and companies with 10 to 100 employees. Before a technology project can take place, the enterprise is visited and evaluated by a country-based technology attaché. Each of the attachés seconded from R&D institutes is responsible for a specific geographical area and acts as broker, or as organizer, animator, or coach, in the innovation process of SMEs. The attachés have an active program of visits to companies and are normally the companies' first contact with the TEFT program. The attachés are in close touch with what is going on in Norwegian technological research institutes, and they put companies in contact with scientists, who will carry out the technology project in close cooperation with the company. The TEFT program pays 75 percent of project costs, while the remaining 25 percent is covered by the company itself. Financial support from TEFT will usually come to between 4,000 and 13,000 euros.

Through these instruments, TEFT is able to help lower barriers to cooperation between national R&D institutes and SMEs, with the attachés acting as brokers; in that way TEFT can encourage SMEs to use R&D institutes and to strengthen industry-science relations.

Sources: European Commission 2002; OECD 2004.

Ways of Encouraging the Establishment of Industrial Clusters

International experience has shown that there are four elementary factors that help bring about a successful industrial cluster for innovative activity: highly skilled technical labor, managerial labor, firm-formation and firm-building capabilities, and connection to markets (Bresnahan, Gambardella, and Saxenian 2001). The government's role is to create the conditions for market-driven clusters to emerge and thrive (see box 1.10).

The governments' focus should be to (1) strengthen the physical and information infrastructures to support the efficient movement of people,

---

Box 1.10 *Ingredients for the Success of Bangalore as an Innovative Cluster*

Bangalore started as a local cluster focused on aeronautics. It slowly grew into IT and then into biotechnology. By the end of the 1990s, many multinational companies had established R&D centers in the city. Bangalore has acquired many of the necessary ingredients to gain such status: good educational institutions, a critical mass of innovative companies, an entrepreneurial culture, and the presence of venture capital.

As they evolve, the focus of clusters moves from production to innovation. This transformation occurs when a threshold number of innovative entrepreneurs exist. As global competition increases, local clusters are becoming crucial for enhancing competitiveness. Clustering and dense interfirm networks provide advantages for firms of all sizes.

Clusters represent a new way of thinking about national, state, and local economies, which imply new roles for government, companies, and other institutions in enhancing competitiveness. The evolution of Bangalore indicates that public policy must focus on designing a set of enabling instruments that must be implemented at the regional and local levels. In the early phases of a cluster, the policy focus has to be on supplying manpower and improving infrastructure. In the growth phase, the focus must be on supporting entrepreneurship, networking, and innovation. In later phases, availability of venture capital is crucial. Local institutions and government must play an increasing role in a cluster if it is to become truly innovative.

Sources: ATIP 2003; Bowonder 2002.

---

goods, and services; (2) fund and establish research and industrial parks and business incubators that encourage innovation-based competition; (3) encourage business investment in R&D and university-industry collaboration through appropriate fiscal incentives; and (4) provide an innovation-friendly set of policies for human resources, intellectual property protection, antitrust, employment, and so forth.

Other Policies

Other policies that are also important for improving innovation performance include sound rule of law for maintaining effective competition in the markets, good governance and accountability, stable macroeconomic policy, a sound financial system, openness to external ideas and free trade, and strong focus on encouraging domestic competition and tapping into the growing global knowledge and technology stocks.

Note

1. There are two well-known Schumpeterian hypotheses regarding the determinants of innovation. The first one deals with the relationship between innovation and monopoly power and stresses the idea that a concentrated market structure boosts innovative activity. The second is concerned with the relationship between firm size and the attitude with regard to investing in innovative activities. In Schumpeter's (1942) view, monopolists have the possibility to attract more qualified scientists and technicians and have, in general, fewer financial constraints. R&D investments are characterized by a lower probability of success than investments in physical capital; in contrast, their potential revenues are usually very high (Scherer, Harhoff, and Kukies 2000). Therefore, R&D is more likely to be performed by firms able to bear risky projects and having the possibility to protect and finance their investments. Firms can use their current market power to obtain resources that can be devoted to R&D. The eventual output of this process allows firms to preserve their market power, earn extra profits that reward the original R&D investment, and make it possible to continue the innovation process.

References

Aghion, Philippe, Nicholas Bloom, Richard Blundell, Rachel Griffith, and Peter Howitt. 2002. "Competition and Innovation: An Inverted U Relationship." NBER Working Paper 9269, National Bureau of Economic Research, Cambridge, MA.

Aghion, P., and R. Griffith. 2005. *Competition and Growth: Reconciling Theory and Evidence (Zeuthen Lectures)*. Cambridge, MA: MIT Press.

Armas, E., I. Goldberg, A. Jaffe, T. Muller, J. Sunderland, and M. Trajtenberg. 2006. "Public Financial Support for Commercial Innovation." Europe and Central Asia Chief Economist's Regional Working Paper Series, vol. 1, no. 1, World Bank, Washington, DC.

Arrow, K. J. 1962. "Economic Welfare and the Allocation of Resources for Inventions." In *The Rate and Direction of Inventive Activity: Economic and Social Factors*, ed. R. R. Nelson. Princeton, NJ: Princeton University Press.

ATIP (Asian Technology Information Program). 2003. "Indian Contract Research." ATIP report, October 7, Albuquerque, NM.

Auerswald, Philip E., and L. M. Branscomb. 2003. "Valleys of Death and Darwinian Seas: Financing the Invention to Innovation Transition in the United States." *Journal of Technology Transfer* 28 (3–4): 227–39.

Bar-Shalom, A., R. Cook-Deegan. 2002. "Patents and Innovation in Cancer Therapeutics: Lessons from CellPro." *The Milbank Quarterly* 80 (4): 637–76.

Beije, P. 2000. *Technological Change in the Modern Economy*. Cheltenham, U.K.: Edward Elgar.

Bowonder, B. 2002. "Evolution of Bangalore as an Innovative Cluster." *Viewpoints.* September 8. Center for International Development, Harvard University, Cambridge, MA.

Bresnahan, Timothy, Alfonso Gambardella, ed. 2004. *Building High-Tech Clusters: Silicon Valley and Beyond.* Cambridge, U.K.: Cambridge University Press.

Bresnahan, T., A. Gambardella, and A. L. Saxenian. 2001. "Old Economy Inputs for New Economy Outcomes: Cluster Formation in the New Silicon Valleys." *Industrial and Corporate Change* 10 (4): 835–60.

Burrone, E., and G. S. Jaiya. 2004. "Intellectual Property (IP) Rights and Innovation in Small and Medium-sized Enterprises." Background document to "The Second OECD Ministerial Conference for Small and Medium-Sized Enterprises," Istanbul, June 3–5,

Chen, D., and Dahlman, C. 2004. "The Knowledge Economy, the KAM Methodology and World Bank Operations." Working Paper 37256, World Bank Institute, Washington, DC.

Chung, Sungchul, and J. Shu. 2007. "Harnessing the Potential of Science and Technology." In *Korea as a Knowledge Economy,* ed. J. Shu and D. Chen. Washington, DC: Korea Development Institute and World Bank Institute.

Cohen, W. M., R. R. Nelson, and J. P. Walsh. 2000. "Protecting Their Intellectual Assets: Appropriability Conditions and Why U.S. Manufacturing Firms Patent (or Not)." NBER Working Paper 7552, National Bureau of Economic Research, Cambridge, MA.

Crespi, F. 2004. "Notes on the Determinants of Innovation: A Multi-Perspective Analysis." Social Science Research Network Electronic Paper Collection. http://ssrn.com/abstract=524503.

EC (European Commission). 2002. "Directory of Measures in Favor of Entrepreneurship and Competitiveness." EC Enterprise and Industry Department. Brussels. http://ec.europa.eu/enterprise/enterprise_policy/charter_directory/en/technology/norway.htm.

Economist. 2007. "Technology in India and China: Running Fast." Special report, November 8. http://www.economist.com/specialreports/displaystory.cfm?story_id=10053169.

Edquist, C., and S. Jacobsson. 1988. *Flexible Automation: The Global Diffusion of New Technology in the Engineering Industry.* Oxford, U.K.: Basil Blackwell.

Goldberg, Itzhak, Manuel Trajtenberg, Adam Jaffe, Thomas Muller, Julie Sunderland, and Enrique Blanco Armas. 2006. "Public Financial Support for Commercial Innovation." Chief Economist's Regional Working Paper Series, Vol. 1, No. 1. ECSPF (Finance and Private Sector Development Sector, European and Central Asia Region), World Bank, Washington, DC.

Greenstein, S., and G. Ramey. 1998. "Market Structure, Innovation and Vertical Product Differentiation." *International Journal of Industrial Organization* 16: 285–311.

Griliches, Z. 1986. "Productivity, R&D, and Basic Research at the Firm Level in the 1970s." *American Economic Review* 76: 141–54.

Hashmi, A. R., and J. V. Biesebroeck. 2006. "Competition and Innovation: A Dynamic Analysis of the US Automobile Industry." Paper presented at the 40th meeting of the Canadian Economics Association held at Concordia University, Montreal, Quebec, May 25–28.

Hong, Y. S. 2005. "Evolution of the Korean National Innovation System: Towards an Integrated Model." In *Governance of Innovation Systems*. Vol. 2: *Case Studies in Innovation Policy*. Paris: OECD.

Krugman, P. R. (1994). "Competitiveness: A Dangerous Obsession." *Foreign Affairs* 73 (2): 28–44.

———. 1996. "Making Sense of the Competitiveness Debate." *Oxford Review of Economic Policy* 12 (3): 17–25.

Levin, R. C., A. K. Klevorick, P. R. Nelson, and S. G. Winter. 1987. "Appropriating the Returns from Industrial Research and Development." Brookings Papers on Economic Activity, 1987: 3. Brookings Institution, Washington, DC.

Lipsey, R., and K. Carlaw. 1998. *A Structuralist Assessment of Technology Policies—Taking Schumpeter Seriously on Policy*. Industry Canada Research Publications Program, Ottawa.

Maleki, E. 1991. *Technology and Economic Development*. New York: John Wiley.

Maskus, K. 2000. *Intellectual Property Rights in the Global Economy*. Washington, DC: Institute for International Economics.

Mehta, D., and B. Joshi. 2002. "Entrepreneurial Innovations in Gujarat" *AI & Society* 16 (1): 100–11.

Morck R., and B. Yeung. 2001. *The Economic Determinants of Innovation*. Ottawa: Industry Canada Research Publications Program.

Nelson, R., ed. 1993. *National Innovation Systems: A Comparative Analysis*. New York: Oxford University Press.

Nuffield Council on Bioethics. 2002. "The Ethics of Patenting DNA: A Discussion Paper." Nuffield Council on Bioethics, London.

OECD (Organisation for Economic Co-operation and Development). 1997. *National Innovation Systems*. Paris: OECD.

———. 2000. *Mobilizing Human Resources for Innovation*. Paris: OECD.

———. 2002. *Science, Technology and Industry Outlook*. Paris: OECD.

————. 2003. *Genetic Inventions, IPRs and Licensing Practices: Evidence and Policies.* Paris: OECD.

————. 2004. *Global Knowledge Flows and Economic Development.* Paris: OECD.

————. 2005. *Governance of Innovation Systems.* Vol. 1: *Synthesis Report.* Paris: OECD.

Porter, M. E., J. L. Furman, and S. Stern. 2000. "The Drivers of National Innovative Capacity: Implications for Spain and Latin America." Harvard Business School, Boston, Massachusetts.

————. 2002. "The Determinants of National Innovative Capacity." *Research Policy* 31: 899–933.

Schacht, H. W. 2000. "Industrial Competitiveness and Technological Advancement: Debate over Government Policy." CRS (Congressional Research Service) Report to the U.S. Congress. Order code: RL33528.

Scherer, Frederic M., Dietmar Harhoff, and Joerg Kukies. 2000, "Uncertainty and the Size Distribution of Rewards from Technological Innovation." *Journal of Evolutionary Economics* 10 (1/2): 175–200.

Schumpeter, Joseph A. 1942/1994. *Capitalism, Socialism, and Democracy.* New York and Oxford, UK: Routledge.

————. 1912/1961. *The Theory of Economic Development: An Inquiry into Profits, Capital, Credit, Interest and the Business Cycle.* Trans. Redvers Opie. Cambridge, MA: Harvard University Press.

Shapira, P., and S. Rosenfeld. 1996. "An Overview of Technology Diffusion Policies and Programs to Enhance the Technological Absorptive Capabilities of Small and Medium Enterprises." Background paper prepared for OECD, Paris.

Sherwood, R. M. 1990. *Strengthening Protection of Intellectual Property Rights and Economic Development.* Boulder, CO: Westview Press.

Tassey, G. 1992. *Technology Infrastructure and Competitive Position.* Norwell, MA: Kluwer Academic Publishers.

Thomas, L. G. 1994. "Britain versus France in Global Pharmaceuticals." *Industrial and Corporate Change* 3 (2): 451–89.

Tyson, L.D'A. 1992. *Who's Bashing Whom: Trade Conflict in High Technology Industries.* Washington, DC: Institute for International Economics.

UNESCO (United Nations Educational, Scientific and Cultural Organization). 2004. *Statistical Yearbook.* Paris: UNESCO.

U.S. FTC (Federal Trade Commission). 1996. "Anticipating the 21st Century: Consumer Protection Policy in the New High-Tech, Global Marketplace." Staff report, U.S. FTC, Washington, DC.

————. 2003. "To Promote Innovation: The Proper Balance of Competition and Patent Law and Policy." Staff Report, U.S. FTC, Washington, DC.

Willig, R. 2006. "Innovation, Growth and Competition." Presentation at the World Bank Knowledge Economy Forum V, Prague, March.

WIPO (World Intellectual Property Organization). 2003. "Intellectual Property: A Power Tool for Economic Growth." Geneva.

World Bank. 2001. China and the Knowledge Economy: Seizing the 21st Century. World Bank Institute Development Study, World Bank.

————. 2007a. Knowledge Assessment Methodology (KAM) database. Knowledge for Development Program, World Bank, Washington, DC.

————. 2007b. *World Development Indicators*. Washington, DC: World Bank.

World Economic Forum. 2008. *Global Competitiveness Report 2007–2008*. Geneva: World Economic Forum. http://www.gcr.weforum.org/.

Yoo, JaeHoon. 2007. "Financing Innovation." *Viewpoint* No. 315 (March), World Bank, Washington, DC.

Zeng, Douglas Zhihua. 2007. "Promoting SME Innovation in China." Background study for *China: Promoting Enterprise-led Innovation* (Economic Sector Work), EASFP (East Asia and Pacific Region, Finance and Private Sector Development Unit), World Bank, Washington, DC.

————, ed. 2008. *Knowledge, Technology, and Cluster-Based Growth in Africa*. Washington, DC: World Bank.

Zeng, Douglas Zhihua, and S. Wang. 2007. "China and the Knowledge Economy: Challenges and Opportunities." Policy Research Working Paper 4223, World Bank, Washington, DC.

PART II

Innovation and Financial Development

2

Financial Innovation and Economic Performance

Shahid Yusuf

Financial innovation has supported and stimulated commercial activities and industrial development over the centuries. This history is described in a handsomely illustrated volume by Goetzmann and Rouwenhorst (2005),[1] and some of the empirical evidence on the role of financial intermediation can be found in Goldsmith (1969). Since the 1970s and 1980s, however, the contribution of the financial sector to the growth of productivity in the industrial countries appears to be rising. This is the outcome of an easing of financial controls, which started in the United States, and the subsequent interaction among financial innovations that leveraged advances in the collection and analysis of market information, institutional changes, and in global financial integration.[2] Now it is the turn of the industrializing economies to enlarge the benefits from financial deepening—and parallel efforts to promote institutional reforms—and from the opening of their markets to international flows of capital.

A few indicators can help to provide a sense of the increasing contribution of the financial sector. Between 1950 and 2005, the sector's share in the U.S. economy rose from 2 percent to 8 percent, and business debt as a share of GDP more than doubled, from 30 percent to 65 percent. Moreover, the contribution of greater financial scale and efficiency to growth is buttressed by research on the United States and cross-country studies.[3] Greenwood, Sanchez, and Wang (2007) estimated that if all countries were to adopt the "best financial practice," global output would rise by 21 percent. They further noted that variations in financial development among countries explain 19 percent of the differences in GDP among countries. Reforms introduced by industrializing nations are accelerating the diffusion of financial technologies, as is globalization, especially via flows of equity capital and foreign direct investment (FDI) in the banking system.

Productivity gains are greatest if the opening of capital markets occurs once the banking system has acquired firm foundations, legal and institutional developments have passed certain thresholds, and institutional reforms have acquired a momentum grounded in widely perceived advantages (Chinn and Ito 2005; Kose et al. 2006). Countries such as China that are entering a critical phase of financial development and integration with the global market are now poised to reap the economic advantages that an innovative, diversified, and effectively regulated financial system can provide.

The purpose of this chapter is twofold. It first briefly discusses the contribution of key financial innovations, and examines the role of information and communication technology in accelerating the pace of innovation.[4] Although innovation in services is at least as important as it is in manufacturing, the latter has tended to receive far more attention than the former (de Vries 2006). The second part of the chapter comments on the status of financial change in China, the speed at which new financial instruments are being introduced into the Chinese financial system, and the areas where catch-up will have fruitful consequences.

What Financial Innovation Does

Financial markets contribute to the efficient functioning of the economy by intermediating the allocation of resources among numerous lenders and borrowers. They help mobilize financial savings and channel them into economically rewarding investments. And financial markets permit a redistribution of resources, both spatially and over time, thereby permitting a separation of income flows and spending. Together, these activities help raise an economy's level of output. Whether financial intermediation causes growth or follows in the wake of growth is still debated. The relationship is bidirectional, and researchers generally find what they go looking for.[5] It is not an issue that econometric tests are ever going to convincingly resolve, but more sophisticated tests have strengthened the case for finance as a determinant of growth (Levine 2005).

Financial innovation often goes hand in hand with the strengthening of legal institutions; for example, safeguarding property rights can have an independent and positive influence on growth by improving allocative efficiency and access to finance (Beck, Demirgüç-Kunt, and Levine 2003, 2005; Claessens and Laeven 2003). Relatedly, managing financial development and ensuring that market participants are not exposed to excessive and destabilizing levels of risk also requires the building of dependable regulatory institutions, a process that must be complemented by measures aimed at improving the technical skills of financial institutions, strengthening their governance structures, and increasing both the awareness of fiduciary responsibilities and the capacity to enforce them (Claessens 2006). The subprime mortgage crisis of 2007–08, and its fallout, has underscored the importance of all these points (Dooley, Folkerts-Landau, and Garber 2008; Morris 2008).

Financial innovations can promote economic activities by making it easier to raise and access capital. First, innovations can provide more convenient and higher-yielding instruments for investors and borrowers. Derivatives permit a tailoring of contracts to suit individual preferences.[6] Together with credit scoring and securitization, they also increase the access of small firms to credit (Bofondi and Lotti 2006). Securitization first started in the United States during the 1970s when lenders began pooling and repackaging mortgages and selling them as securities (Econbrowser 2008; Rosen 2007). It has since spread far and wide, embracing many other kinds of instruments. For example, business entities in developing countries are now raising funds by securitizing future receivables and selling them to an "offshore special purpose vehicle (SPV) which issues debt instruments" (Ketkar and Ratha 2001, 46). In the United States for example, securitization now accounts for 53 percent of nonfinancial debts, compared with 28 percent in 1980. Credit scoring and the Internet permit the geographical expansion of banking by facilitating long-distance transactions with customers (Berger and DeYoung 2006).

Second, by assisting with the collection and analysis of information, innovations can reduce frictions and transaction costs, improving the efficiency of financial intermediation (Tufano 2003). Third, they can make it easier for lenders to monitor the use of their funds by entrepreneurs. Fourth, innovations are instrumental in transforming the riskiness of instruments. Credit derivatives, for example, reduce capital costs by decomposing the various risks associated with a security (such as the yield, holding period, and repayment) and repackaging the underlying value into new and differentiated instruments (see *Economist* 2007). .

Recent history shows that innovations can take a number of forms.[7] The most common are new products such as adjustable-rate mortgages. Innovation also widens the range of services that are offered, an important addition being Internet banking. Electronic record keeping and credit-scoring techniques have added to the menu of production processes available to financial institutions. Moreover, a raft of organizational innovations has given rise to new entities such as electronic exchanges and Internet-only banks.

How IT Boosts Innovation

The past 15 years have witnessed a remarkable acceleration in all forms of innovation, in large part because of advances in information and communication technology (ICT, but usually referred to as IT).[8] This has touched many parts of the economy, but arguably, several services industries have benefited more than others, finance being among the most fortunate, followed by retailing, logistics, and a number of the creative industries such as movie making and video games (Aglietta and Breton 2001; Solow 2001; Yusuf and Nabeshima 2006).

Innovation has frequently been spurred by cheaper and more reliable information, transaction costs, and the presence of information asymmetry. For example, the greater availability of accounting information in the late-19th century was responsible for the spread of unsecured income bonds that paid interest only when the firm issuing them registered an accounting profit (Tufano 2003).

IT and digital technologies have opened the door to a host of new ways of conducting financial business. Probably the most striking development is the enormous increase in data that IT has provided. This would be a mixed blessing if the supply were not matched by a vast expansion of computing power. It is this computing capability that makes it possible to extract usable information from the deepening pool of data and underlies the surge in innovations. Typically it is those banking and nonfinancial firms that have invested in skills to fully harness the IT revolution that have benefited most (Lin 2007).

Financial engineering predates the IT revolution, but it is undoubtedly the case that the application of value-at-risk (VaR) concepts to finance and the adoption of the models of capital asset pricing, options, and risk assessment would simply not have happened across the world in the absence of increasingly powerful computers. Transaction costs motivated the adoption of the analytic framework and the option pricing model devised by Black, Scholes, and Merton, which provides the foundations for new valuation techniques, and a multitude of derivatives. But the widespread applications of Black-Scholes is inconceivable without sophisticated computer hardware and equally sophisticated software products, which bring the power of highly esoteric theorems to the desktop of the average financial analyst (Bernstein 2007; Tufano 2003). Furthermore, it is the combination of hardware and ingenious programs that permits the numerous innovations in complex financial products, structured and customized for myriad and highly specific market niches (Bernstein 2007). The subprime mortgage crisis has shown that the complexity of some of the instruments has proved to be a mixed blessing by making it harder to assess the value of the derivative products at any time and by concealing the risk exposure of the entities holding these instruments.

There are two additional and interlinked avenues through which IT has induced innovation. One is financial globalization; the second is deverticalization and outsourcing. First, there can be no doubt that because of advances in communications technology and the Internet, financial globalization has progressed much further than other forms of global integration. This has generated vast opportunities for financial intermediaries. By 2006, cross-border investments had risen to US$6 trillion. Spearheading this process are multinational corporations and, more recently, hedge funds (controlling US$2.7 trillion in 2007), sovereign wealth funds (controlling US$3.3 trillion in 2007), and private equity funds. Increased openness, the perception of interdependence, and the greater need for coordination and conformity on many fronts are also responsible for the greater adherence to sound macroeconomic policies, to steady gains in the quality of financial market regulation, and to the significance now attached to financial transparency and governance (Kose et al. 2006). These have given added impetus to globalization.

Second, IT has spurred the subdivision and outsourcing of activities that financial institutions once performed under one roof. This process innovation has seen many back-office and other transactions transferred to specialized providers who can fully exploit economies of scale and of learning by doing. With IT, financial intermediaries are able to slash their costs by delayering the organization and transferring activities such as accounting, bookkeeping, legal, human resource management, call centers, and even risk assessment to the lowest-cost and most efficient providers throughout the world (McKinsey Global Institute 2003). By one estimate, between 2004 and 2009, US$356 billion in U.S. financial services will be sent offshore (see Bank for International Settlements 2005). And aside from routine activities, the design of sophisticated financial instruments, as well as the writing of code to harness these instruments, can also be more expeditiously outsourced to academia or experts, wherever they might be.[9]

Payoffs of Innovation

There is no denying the volume and variety of financial innovations, but what is the payoff? The answer can be provided at three levels: evidence from cross-country research that has measured the impact of financial development on aspects of economic performance, evidence from studies of the sources of growth, and findings from the empirical analysis of financial businesses.

First, financial depth and the sophistication of various financial markets are associated with more rapid growth and with the better performance of firms. This is documented by a considerable literature emphasizing the role that improvements in the financial system can make by reducing transaction costs, increasing allocative efficiency, and making credit more easily available to firms (helped by credit scoring and securitization)—all of which can be growth enhancing (Levine 1997, 2005; Levine and Zervos 1998). Financial depth and innovation can strengthen corporate accountability. These developments can also enable small and medium firms to satisfy their capital needs (Mishkin and Strahan 1999).

Second, findings, mainly from the United States and Europe, highlight the contribution of services to GDP growth. The U.S. economy has grown at an unusually rapid pace since the late 1990s. In large part, this is the result of continuous and substantial gains in total factor productivity averaging 2.5 percent and more per year. At the root of this are the productivity gains in the manufacture of IT products stimulated by investment in IT—which rose from 2 percent of GDP in 1980 to 3 percent in 1990 to almost 6 percent in 2000—and productivity gains in major segments of the economy from the adoption of IT technology (Jorgenson 2005). Among these, financial services and securities trading achieved some of the highest rates of productivity increase. Relative to the United States, Europe has lagged behind, in part because services industries have been slower to assimilate IT and capitalize on the opportunities it offers for innovation

(Jorgenson et al. 2007). One line of thinking is that financial innovations were critical to sustaining U.S. productivity growth because they facilitated the entry of new and innovative firms, encouraged "creative destruction," channeled resources into the most profitable uses, and helped attract capital from overseas. Financial innovation was by no means the major driver of growth; however, in conjunction with IT and other subsectors, it appears to have made a substantial contribution.

The third piece of evidence comes from studies of specific innovations and of specific categories of financial firms. Automated teller machines, smart cards, and automated clearing house (ACH) network transfers have all served to significantly lower transaction costs (Tufano 2003). Innovations by the mortgage industry—tailoring mortgages (fixed, variable, and adjustable rate) for different types and risk classes of borrowers and securitizing those mortgages, creating secondary markets, and designing securities to suit the preferences of different investors—have enormously increased the availability of mortgage financing and widened home ownership (Gerardi, Rosen, and Willen 2007). With the help of IT, firms have introduced open initial public offerings (without underwriting) and an assortment of new methods for assembling portfolios and executing security transactions (Tufano 2003).

IT plus a variety of regulations on interest payments were behind the introduction of other profitable innovations such as cash management accounts, money market accounts, mutual funds, and sweep accounts for bank deposits (Frame and White 2004). The quantum jump in computing power, combined with intellectual innovations, has also resulted in techniques that have yielded high returns. For example, VaR analysis and stress testing (Cihak 2007) of portfolios have been found to be effective and have been widely adopted (Frame and White 2004).

Who the Innovators Are

Clearly, the scale of innovative activity reflects the attractiveness of new financial instruments and practices. But who is responsible for the innovations? An interesting study of financial innovations in the United States between 1990 and 2002 found that, on average, smaller and less profitable financial firms are more innovative (Lerner 2006). This contributes to their future profitability. To a lesser degree, so also are older and larger firms operating in areas where innovation is flourishing. Large banks and fast-growing banks with branch networks that allow them to exploit scale economies have usually taken the lead in widely adopting credit scoring, ATMs, Internet services, and the issuance of new types of securities (Frame and White 2004; Tufano 2003). The concentration of market share in the banking sector has also induced banking relationships that provide readier access to credit, although greater banking competition allows for easier entry of firms (Cetorelli and Gambera 2001; Deidda and Fattouh 2005). As with innovation in other fields, neighborhood effects and clustering seems to influence innovation positively. A tiny number of cities in the United States—New York and Chicago being preeminent—are responsible for most of the financial innovations. This pattern is evident in Europe and is being repeated in East Asia (Cassis 2007). Innovation is very likely to be a phenomenon that radiates outward from the major urban centers in China.

How Innovation Will Persist

The Nobel Prize winner Robert Merton sees the financial system as the "engine" driving the real economy. He ascribes some of the dynamism of the U.S. economy as well as that of leading industrialized countries to the creation of national mortgage markets, markets for financial derivatives, the mutual fund industry, and markets for a variety of securities that provide liquidity and help finance long-lived capital assets. Merton, according to Bernstein (2007), is convinced that "innovations developed by profit seeking institutions, like mutual funds and insurance companies can mitigate and even overcome the behavioral anomalies and market inefficiencies created by individual investors in the real world" (p. 50). Merton is also of the view "that the most fruitful source for continuing the spiral of financial innovation will develop primarily from the valuation of options—or more precisely, contingent claims" (p. 52).

There are several reasons for believing that the pace of financial innovation may not slacken soon. First, the potential of IT, digital technologies, and globalization to spawn financial innovation is far from exhausted. As with electric power and the internal combustion engine in the early 20th century, the technology windfall may continue to accrue for another two to three decades.

Second, the financial industry is concentrated in some of the world's most creative cities and benefits from access to the best talent, links with the leading academic centers, and interaction with the fastest-growing new industries. Locational advantages and the talent pool will remain major assets for the financial sector. Third, the greater protection given to intellectual property, and to financial innovations in particular, could stimulate innovation in industrialized and industrializing countries.[10]

And fourth, the aging of the populations in industrial and some industrializing economies is opening the door to new instruments. Pension funds seek long-duration bonds that are inflation indexed, especially in view of uncertainty regarding future price movements. Households require annuities to meet the needs of long-term care. There is, in addition, an emerging demand for reverse mortgages, survivor and mortality bonds, and mortality derivative securitization (Mitchell et al. 2006). These instruments represent only the first wave of innovation and undoubtedly the ingenuity of the financial industry will bring forth others. Much like earlier innovations, these will help to distribute risk more evenly, enhance the liquidity of assets, and assist with the intertemporal distribution of consumption for societies in which more and more people are in the upper age brackets.

China's Financial System

As global market integration accelerates the transfer of financial innovations to industrializing countries, the potential for deriving productivity gains from these is greatest in the fast-growing East Asian economies, especially China.

In 2005, China's financial sector accounted for 3.4 percent of GDP, or approximately 9.0 percent of the output of tertiary industries (National Bureau of Statistics 2007). Although growth of total factor productivity during 1993–2004 has averaged 4 percent annually—almost 41 percent of the aggregate growth of GDP (Bosworth and Collins 2007)—the contribution of the financial sector and other leading services industries to productivity is miniscule.

By international standards, China's economy has achieved substantial financial depth. Starting from low levels in 1980, financial ratios have risen very quickly (Lane and Schmukler 2006). Between 1991 and 2004 alone, the ratio of bank credit to GDP doubled, and deposits to GDP rose by a factor of three.[11] Total banking assets by end-2006 amounted to 44 trillion yuan, almost 210 percent of GDP. They are now much higher than those of other East and South Asian comparators and of the G-7 countries. However, the financial sector remains relatively inefficient; its productivity is well behind that of the industrial countries. According to a study by the McKinsey Global Institute (2006), financial development could enable China to raise its GDP by US$320 billion.[12]

China's banking system accounts for the majority of financial assets: 72 percent of all financial assets in 2004,[13] versus 45 percent in Thailand, 43 percent in India, and 33 percent in the Republic of Korea. The majority of the banking sector's loans are still to state and collectively owned enterprises (73 percent), and loans are a source of most of the banks' income.[14] The banks, moreover, have been slow to introduce organizational changes, adopt IT, and outsource and raise the skill levels of staff, which constrains their ability to introduce some of the innovations and risk assessment techniques described above (Farrell, Lund, and Morin 2006). Consequently, their operating costs are high. The banking system is hampered further by distortions arising from interest rate controls.

The past decade has witnessed considerable development of the bond, mortgage, and consumer credit markets. This has permitted a sixfold increase to US$103 billion (in 2006) in the funding mobilized from nonbank sources. In the first nine months of 2006, US$27 billion was raised through the issuance of short-term bonds and US$13 billion was raised by floating of long-term bonds (*Economist* 2006).[15] This is a promising surge; however, there is much ground to be covered both with respect to the mix of instruments and the expertise to support them. The corporate bond market remains small. In 2006, it was equal to 13 percent of GDP, and bonds accounted for 1 percent of corporate debt, whereas in other economies it can be as high as 60 percent (Farrell, Lund, and Morin 2006). The same applies to the mortgage market, now Asia's largest, but accounting for just one-tenth of the GDP. The growth in the share of this market, which is a key to China's urbanization, will depend on adopting innovations such as securitization and institutions that help to build a secondary mortgage market (*Shanghai Daily* 2006).

The stock market, after a steep ascent in 2006, had achieved a capitalization of US$1.4 trillion by the end of the year (about one-tenth of that of the United States), with 1,400 listed companies.[16] However, its financial role is circumscribed because tradable shares represent only about one-third of stock market capitalization, and investors are deterred by concerns regarding equity pricing as well as by the adequacy and transparency of corporate governance in listed firms.[17]

China's larger firms and municipalities have little difficulty obtaining credit, especially from banks. Smaller firms and households have less access, although the situation is improving rapidly. But for small firms, the strict rules that allow banks to accept only real estate as collateral present major hurdles. Medium-size firms cannot tap the bond market because of regulatory constraints, a lengthy approval process, and guarantee- and cost-related issues. Raising funds from the stock market is equally problematic and costly. Moreover, there is considerable variation in the ratios of bank credit to provincial output, with the poorer and slower-growing provinces having higher ratios.

China's financial sector (for example, the securities market) also is still weakly integrated with the global system. This shelters it from some risks but also shelters it from competition and reduces technology transfer, the propensity to innovate, and the option to raise funding from regional sources (for example, through the issuance of bonds).[18] The limited extent of global integration means, in addition, that economies of scale, scope, and specialization might be accruing more slowly than would be the case otherwise. This affects the role China's financial centers can play. Although Hong Kong, China, ranks fifth among the international financial centers, Shanghai is at 32nd place, far behind Singapore and Seoul, which are ranked sixth and ninth, respectively (see table 2.1).

Finally, while legal reform is proceeding with a focus on the Company Law, the Securities Law, and the Securities Investment Fund Law, the pace may have to quicken in order for legal institutions to more fully assist the transformation of the financial system (Liu 2007). In short, stimulating innovation would yield large benefits. But a deepening of mortgage finance securitization, the introduction of derivatives, foreign exchange trading, and a greater role for hedge funds will call for a substantial increase in skills, the effective use of IT, a strengthening of managerial capacity, and the spur provided by increased competition. In addition, a move toward greater innovation should be paralleled by a strengthening of regulatory and monitoring capabilities to buffer financial markets against shocks (Roubini 2008).

A Summing Up

In summarizing the main messages of this chapter, four points deserve to be underlined. First, the financial systems of the leading industrial countries are an unusually prolific source of innovations, and these are now spreading rapidly to industrializing countries. Second, as a result of the IT revolution, great advances in computing technologies, and globalization, the productivity of the financial sector has risen rapidly over the past decade, especially in the United States and now, potentially, in industrializing East Asia. Third, financial innovations, which enhance the efficiency of resource allocation, cut down transaction costs, promote liquidity, and better serve the requirements of borrowers and lenders, are making a significant difference to allocative efficiency and total factor productivity,

Table 2.1 Ranking of the Top 50 Worldwide Centers of Commerce, 2007

Rank	City	Index values	Legal and political framework	Economic stability	Ease of doing business	Financial flow	Business center	Knowledge creation and information flow
1	London	77.79	84.11	93.54	87.87	82.86	71.75	52.72
2	New York	73.80	86.60	90.64	87.36	68.58	62.13	61.55
3	Tokyo	68.09	83.88	88.03	85.52	53.39	59.54	55.94
5	Hong Kong	62.32	77.57	85.94	87.07	38.06	71.89	27.31
6	Singapore	61.95	88.12	91.64	88.24	32.59	66.16	28.78
9	Seoul	60.70	74.63	86.42	73.83	53.00	51.37	42.95
32	Shanghai	50.33	68.63	79.40	64.46	38.30	52.09	17.16
36	Bangkok	47.96	68.01	88.13	68.71	23.21	49.50	16.27
37	Dubai	46.61	78.04	75.63	73.76	23.04	48.82	4.24
45	Mumbai	42.70	50.76	90.51	55.09	38.71	30.34	14.73
48	Sao Paulo	41.14	49.83	67.83	63.06	31.43	34.23	14.50

Source: MasterCard Worldwide 2007.

with positive consequences for economic growth. Aging populations in industrial countries are beginning to unleash a new wave of innovation. Fourth, China has yet to exploit the full potential of financial innovation. Given the financial depth China has achieved and the size of its banking system, technological catch-up with a suitable upgrading of regulatory institutions should be an integral part of China's strategy for innovation and growth.[19] This will entail a further opening of its financial markets and could be assisted by greater foreign investment in banks and securities firms. Both the opening of markets and increased investment will transfer technology and stimulate competition.

Notes

1. The volume by Goetzmann and Rouwenhorst describes the old Babylonian loan tablets, the first known financial instruments. It then roams over the early financial landscape of China before exploring financial developments in Europe, starting in the 12th century, and selectively following the trail of innovation in Europe and the United States to the present.

2. Cassis (2007, 248–49) observes that financial innovation since the 1980s was the outcome of three factors: "First, there was monetary instability, due to inflation and to newly floating currencies that, although providing dealers with new earning opportunities, meant that there was a need to hedge against fluctuations in interest rates and in foreign exchange rates. Next, there was the incredible progress made in computing. Finally, there was the application to the markets of what could be called fundamental research, in other words theoretical advances made by a certain number of economists in the fifties and sixties." On financial liberalization in the 1970s and later and current policy issues regarding financial development, see De La Torre, Gozzi, and Schmukler (2007).

3. See for example Levine (2005) and the research presented in Demirgüç-Kunt and Levine (2001).

4. As noted by Lin (2007), the payoff from IT in the form of efficiency and induced innovation is greatest when a financial enterprise makes the complementary investment in human capital so as to help build IT capability.

5. See Chang and Caudill (2005) and Shan (2005). Using value-at-risk analysis, the first finds that finance causes growth, the second detects only a weak relationship.

6. Grinblatt and Longstaff (2000) noted that derivatives reallocate cash flows among investors based on need and permit the reengineering of portfolios so that securities can be held in their most valuable forms.

7. On this see the reviews by Frame and White (2004) and Tufano (2003).

8. See Cohen, DeLong, and Zysman (2000) on the impact of the IT revolution.

9. The risks from outsourcing are manifold, are coming to be recognized, and are tempering the initial enthusiasm. Firms are reappraising the advantages and seeking to maximize gains by reengineering their own processes; taking greater care in selecting and monitoring vendors subject to different regulations; and attempting to ensure confidentiality of information, maintain strategic capabilities, and put in place plans for contingencies (Bank for International Settlements 2005; Federal Reserve Bank of New York 1999; Gottfredson, Puryear, and Phillips 2005).

10. Intellectual property protection for IT and financial innovation can be counterproductive beyond a point, and there are signs in the United States that the avalanche of patents from services industries is triggering a backlash.

11. China's leading banks are now among the world's largest.

12. For instance, a shift from paper-based to an electronic payment system would save US$62 billion (Farrell, Lund, and Morin 2006).

13. Although pawn shops and mutual savings associations have been present in China for centuries, the first banks, called *piao-hao,* emerged in the 19th century, with bankers from Shanxi taking the lead (Goetzmann, Ukhov, and Zhu 2007). The first *piao-hao* was established in Pingyao.

14. Credit card services are expanding but are not yet yielding much income, nor are commissions a major source of revenue.

15. A fairly up-to-date summary of financial developments can be found in Liu (2007).

16. China also has three commodity futures exchanges and a multitier bond market.

17. The legal framework is also an issue that affects Shanghai's ranking in the financial world (see table 2.1). For the recent history of the stock exchanges in Shanghai and Shenzhen, see "Shenzhen Stock Exchange" (Wikipedia 2007b) and "Shanghai Stock Exchange" (2007a). Goetzmann, Ukhov, and Zhu (2007) observed that even though the Shanghai China Merchants Stock Exchange had become the biggest stock exchange in East Asia by 1935, the development of equity financing then (as it is now) was hobbled by "the ineffectiveness of legal protection (e.g., for minority shareholders) and governance structures for enterprises" (294).

18. The Asian Bond Market Initiative represents a first step in this direction. See "China's Path to World-Class Capital Markets" (*Financial Times* 2007) regarding the scope for more competition through the entry of foreign securities firms, which currently can own only a minority stake in Chinese firms.

19. The creation of the China Banking Regulatory Commission in 2003 was an important step. Acquiring sufficient numbers of staff with the appropriate skills is the next big hurdle. Also, regulators must contend with the authority of local bank managers—in what is still a fairly decentralized banking system—and the influence of local officials over the bank managers.

References

Aglietta, Michel, and Regis Breton. 2001. "Financial Systems, Corporate Control and Capital Accumulation." *Economy and Society* 30 (4): 433–66.

Bank for International Settlements. 2005. *Outsourcing in Financial Services.* Basel, Switzerland: The Joint Forum Basel Committee on Banking Supervision.

Beck, T., Asli Demirgüç-Kunt, and Ross Levine. 2003. "Law and Finance: Why Does Legal Origin Matter?" *Journal of Comparative Economics* 31: 653–75.

———. 2005. "Law and Firms' Access to Finance." *American Law and Economics Review* 7: 211–52.

Berger, Allen, and Robert DeYoung. 2006. "Technological Progress and the Geographic Expansion of the Banking Industry." *Journal of Money, Credit, and Banking* 38 (6): 1483–89.

Bernstein, Peter L. 2007. *Capital Ideas Evolving.* Hoboken, NJ: Wiley & Sons.

Bofondi, Marcello, and Francesca Lotti. 2006. "Innovation in the Retail Banking Industry: The Diffusion of Credit Scoring." *Review of Industrial Organization* 28 (4): 343–58.

Bosworth, Barry P., and Susan M. Collins. 2007. "Accounting for Growth: Comparing China and India." NBER Working Paper 12943, National Bureau of Economic Research, Cambridge, MA.

Cassis, Youssef. 2007. *Capitals of Capital: A History of International Financial Centres, 1780–2005.* Cambridge, U.K.: Cambridge University Press.

Cetorelli, Nicola, and Michele Gambera. 2001. "Banking Market Structure, Financial Dependence and Growth: International Evidence from Industry Data." *Journal of Finance* 56 (2): 617–48.

Chang, Tsangyao, and Steven Caudill. 2005. "Financial Development and Economic Growth: The Case of Taiwan." *Applied Economics* 37: 1329–35.

Chinn, Menzie D., and Hiro Ito. 2005. "What Matters for Financial Development? Capital Controls, Institutions, and Interactions." NBER Working Paper 11370, National Bureau of Economic Research, Cambridge, MA.

Cihak, Martin. 2007. "Introduction to Applied Stress Testing." IMF Working Paper WP/07/59, International Monetary Fund, Washington, DC.

Claessens, Stijn. 2006. "Current Challenges in Financial Regulation." Policy Research Working Paper WPS4103, World Bank, Washington, DC.

Claessens, Stijn, and Luc Laeven. 2003. "Financial Development, Property Rights, and Growth." *Journal of Finance* 63 (6): 2401–36.

Cohen, Stephen, J. Bradford DeLong, and John Zysman. 2000. "Tools for Thought: What Is New and Important about the 'E-Conomy.'" Berkeley Roundtable on the International Economy Working Paper 138, University of California, Berkeley. http://repositories.cdlib.org/brie/BRIEWP138.

De La Torre, Augusto, Juan Carlos Gozzi, and Sergio L. Schmukler. 2007. "Financial Development: Maturing and Emerging Policy Issues." *World Bank Research Observer* 22 (1): 67–102.

de Vries, Eric J. 2006. "Innovation in Services in Networks of Organizations and in the Distribution of Services." *Research Policy* 35: 1037–51.

Deidda, Luca, and Bassam Fattouh. 2005. "Concentration in the Banking Industry and Economic Growth." *Macroeconomic Dynamics* 9: 198–219.

Demirgüç-Kunt, Asli, and Ross Levine, eds. 2001. *Financial Structure and Economic Growth.* Cambridge, MA: MIT Press.

Dooley, Michael, P., David Folkerts-Landau, and Peter M. Garber. 2008. "Will Subprime Be a Twin Crisis for the United States?" NBER Working Paper 13978, National Bureau of Economic Research, Cambridge, MA.

Econbrowser. 2008. "Mortgage Securitization." January 11. http://www.econbrowser.com.

Economist. 2006. "Out of the Shadows." December 16.

———. 2007. "Credit Derivatives." April 21.

Farrell, Diana, Susan Lund, and Fabrice Morin. 2006. "How Financial-System Reform Could Benefit China." *The McKinsey Quarterly* (2006 Special Edition): 92–105.

Federal Reserve Bank of New York. 1999. *Outsourcing Financial Services Activities: Industry Practices to Mitigate Risks.* New York: Federal Reserve Bank.

Financial Times. 2007. "China's Path to World-Class Capital Markets." May 14.

Frame, W. Scott, and Lawrence J. White. 2004. "Empirical Studies of Financial Innovation: Lots of Talk, Little Action?" *Journal of Economic Literature* 42: 116–44.

Gerardi, Kristopher, Harvey S. Rosen, and Paul Willen. 2007. "Do Households Benefit from Financial Deregulation and Innovation? The Case of the Mortgage Market." NBER Working Paper 12967, National Bureau of Economic Research, Cambridge, MA.

Goetzmann, William N. and Geert Rouwenhorst (eds.) 2005. *The Origins of Value.* New York: Oxford University Press.

Goetzmann, William N., Andrey D. Ukhov, and Ning Zhu. 2007. "China and the World Financial Markets 1870–1939: Modern Lessons from Historical Globalization." *Economic History Review* 60(2): 267–312.

Goldsmith, Raymond W. 1969. *Financial Structure and Development.* New Haven, CT: Yale University Press.

Gottfredson, Mark, Rudy Puryear, and Stephen Phillips. 2005. "Strategic Sourcing: From Periphery to the Core." *Harvard Business Review* Online. http://www.itmission.com/outsourcing.pdf.

Greenwood, Jeremy, Juan M. Sanchez, and Cheng Wang. 2007. "Financing Development: The Role of Information Costs." NBER Working Paper 13104, National Bureau of Economic Research, Cambridge, MA.

Grinblatt, Mark, and Francis A. Longstaff. June 2000. "Financial Innovation and the Role of Derivative Securities: An Empirical Analysis of the Treasury STRIPS Program." *Journal of Finance* 55 (3): 1415–36.

Jorgenson, Dale W. 2005. "Accounting for Growth in the Information Age." In *The Handbook of Economic Growth*, ed. Philippe Aghion and Steven N. Durlauf, 744–815. Amsterdam, The Netherlands: Elsevier.

Jorgenson, Dale W., S. Ho Mun, Jon D. Samuels, and Kevin J. Stiroh. 2007. "Industry Origins of the American Productivity Resurgence." Harvard University, Cambridge, MA.

Ketkar, Suhas, and Dilip Ratha. 2001. "Securitization of Future Flow Receivables." *Finance and Development* 38 (1). http://www.imf.org/external/pubs/ft/fandd/2001/03/ketkar.htm.

Kose, Ayhan M., Eswar Prasad, Kenneth Rogoff, and Shang-Jin Wei. 2006. "Financial Globalization: A Reappraisal." NBER Working Paper 12484, National Bureau of Economic Research, Cambridge, MA.

Lane, Philip, and Sergio L. Schmukler. 2006. "The International Financial Integration of China and India." CEPR Discussion Paper 5852, Center for Economic Policy Research, London.

Lerner, Josh. 2006. "The New New Financial Thing: The Origins of Financial Innovations." *Journal of Financial Economics* 79: 223–255.

Levine, Ross. June 1997. "Financial Development and Economic Growth: Views and Agenda." *Journal of Economic Literature* 35 (2).

———. 2005. "Finance and Growth: Theory and Evidence." In *The Handbook of Economic Growth*, ed. Philipp Aghion and Steven N. Durlauf, 865–934, Amsterdam, The Netherlands: Elsevier.

Levine, Ross, and Sara Zervos. 1998. "Stock Markets, Banks, and Economic Growth." *American Economic Review* 88: (3): 537–58.

Lin, Bou-Wen. 2007. "Information Technology Capability and Value Creation: Evidence from the US Banking Industry." *Technology in Society* 29: 93–106.

Liu, Lisheng. 2007. "An Overview of China's Financial Markets." In *China's Financial Markets*, ed. Salih N. Neftci and Michelle Yuan Menager-Xu, 1–35. Amsterdam, The Netherlands: Elsevier Academic Press.

MasterCard Worldwide. 2007. *Worldwide Centers of Commerce Index.* http://www.mastercard.com/us/company/en/insights/pdfs/2007/index_2007_us.pdf.

McKinsey Global Institute. 2003. "Offshoring: Is It a Win-Win Game?" *Perspective.* August 2003. http://www.mckinsey.com/mgi/publications/win_win_game. asp.

———. 2006. "Putting China's Capital to Work: The Value of Financial SystemReform."http://www.mckinsey.com/mgi/publications/china_capital/ executive_summary.asp.

Mishkin, Frederic S., and Philip E. Strahan. 1999. "What Will Technology Do to Financial Structure?" NBER Working Paper 6892, National Bureau of Economic Research, Cambridge, MA.

Mitchell, Olivia S., John Piggot, Michael Sherris, and Shaun Yow. 2006. "Financial Innovation for an Aging World." NBER Working Paper 12444, National Bureau of Economic Research, Cambridge, MA.

Morris, Charles R. 2008. *The Trillion Dollar Meltdown.* New York: Public Affairs.

National Bureau of Statistics. 2007. *China Statistical Yearbook 2007.* Beijing: China Statistics Press.

Rosen, Richard J. 2007. "The Role of Securitization in Mortgage Lending." Essays on Issues No. 244, The Federal Reserve Bank of Chicago.

Roubini, Nouriel. 2008. "Ten Fundamental Issues in Reforming Financial Regulation and Supervision in a World of Financial Innovation and Globalization." *RGE Monitor,* March 31, 2008.

Shan, Jordan. 2005. "Does Financial Development 'Lead' Economic Growth? A Vector Auto-Regression Appraisal." *Applied Economics* 37: 1353–67.

Shanghai Daily. 2006. "Mortgage Industry Untapped, Says BIS," December 11.

Solow, Robert. 2001. "Information Technology and the Recent Productivity Boom in the U.S." Presented at the National Competitiveness Network Summit, Cambridge, MA, November 2001.

Tufano, Peter. 2003. "Financial Innovation." In *Handbook of the Economics of Finance,* ed. G. M. Constantinides, M. Harris, and R. Stulz. Amsterdam, The Netherlands: Elsevier North-Holland.

Wikipedia. 2007a. "Shanghai Stock Exchange." http://en.wikipedia.org/wiki. Shanghai_Stock_Exchange, last modified on July 6, 2008.

———. 2007b. "Shenzhen Stock Exchange." http://en.wikipedia.org/wiki/Shenzhen_Stock_Exchange, last modified on June 16, 2008.

Yusuf, Shahid, and Kaoru Nabeshima. 2006. *Post Industrial East Asian Cities.* Palo Alto, CA: Stanford University Press.

3

China's Financial Sector Policies

Vivek Arora

This chapter discusses China's experience with financial sector development and some of the challenges that remain, using the international experience as a background. The link with innovation in this chapter is more indirect than it is in the other chapters of this book, because the emphasis is not on policies targeted specifically at innovation, but on the role of the overall financial system—that is, a stable, robust system with sound banks and deep capital markets—in providing a solid foundation from which innovative ideas can be given economic life. China has made significant progress in financial development in recent years, particularly in reforming the banking system and developing capital markets. However, the financial system still favors large state-owned enterprises over smaller firms and private enterprises, and, more generally, much remains to be done to complete bank reforms and to further develop capital markets that can play a larger role in the financial system.

International experience suggests that financial development positively influences economic growth, and that it does so in large measure through its impact on productivity and innovation. A more developed financial system, with competition among banks, sound institutions, and developed capital markets, can help to allocate scarce resources efficiently, ensuring that they are most productively utilized.[1] By offering a variety of financing channels, it can help to lower the cost of finance and increase the likelihood of innovative ideas and projects being financed, including those involving smaller firms. A level playing field across firms in terms of access to finance helps to facilitate competition, which is important for innovation. As Greenspan (2002) noted, "Competition is the facilitator of innovation. And . . . the process by which less-productive capital is displaced with innovative . . . technologies is the driving force of wealth creation." Moreover, new financial products can help to increase people's consumption possibilities and to strengthen their ability to insure themselves against income risks, including those associated with old age.

Financial Development, Innovation, and Growth: International Experience

Experience across a wide range of countries suggests that financial and economic development are positively related (figure 3.1).[2] Countries with high levels of per capita income tend to have financial systems that are highly developed. Equivalently, one can say that countries that are financially well developed tend to have high levels of per capita income. The correlation raises a question of causality: does financial development contribute to economic growth or does growth lead to financial development, perhaps because people in richer countries need or demand higher levels of financial services?

The empirical literature suggests that, although there is a "virtuous circle" between financial development and growth, on balance the causality goes more from financial development to economic development.[3] Put simply, as Joseph Schumpeter (1912/1949) emphasized, finance helps to bring innovative ideas to life so that their economic value can be realized. Growth regressions generally suggest that financial development has a large, positive impact on economic growth (King and Levine 1993; Levine 1997). And sectoral analyses find that industries that depend heavily on external resources for their financing tend to grow more slowly than other industries in countries with less-developed financial markets (Rajan and Zingales 1998).

What is the channel through which financial development influences economic growth? A notable conclusion in the empirical literature is that the positive influence of financial development owes mainly to its impact on productivity rather than on investment and employment (Beck, Levine, and Loayza 2000). One can think of productivity as being driven by innovation. Indeed, as Schumpeter envisioned, finance does seem to support

Figure 3.1 *Relationship of Financial Deepening and Development, 1970–2000*

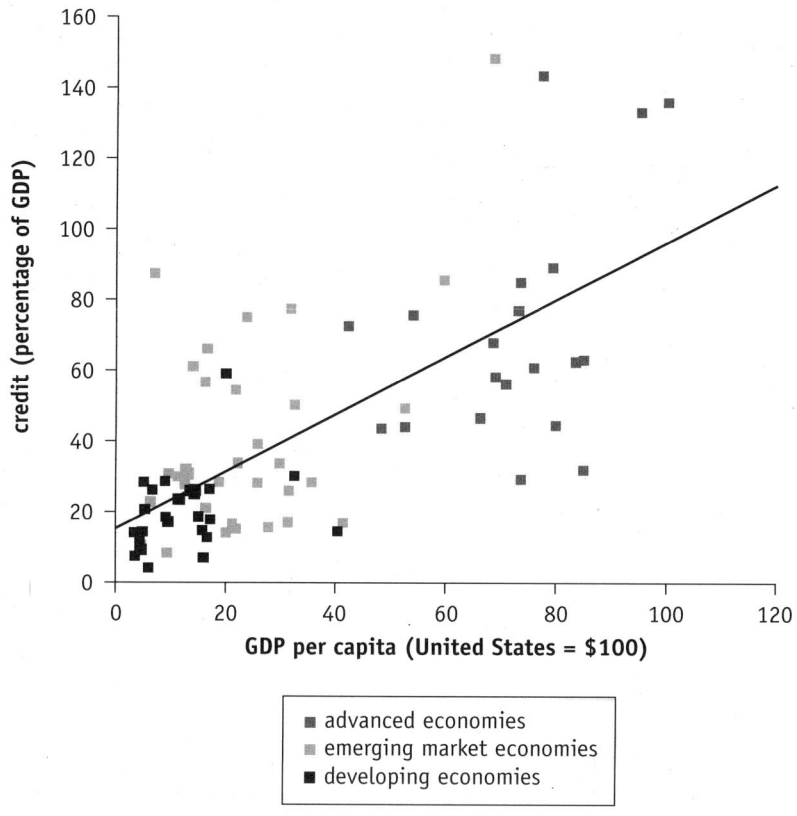

Source: IMF 2004.
Note: Annual average of GDP per capita and bank credit-GDP ratio in selected countries during 1970–2000.

entrepreneurship. The availability of more and better banking services is a first step, but capital markets also contribute positively to economic growth by helping firms diversify their funding sources and improve their access to finance so that they do not have to rely only on banks or their own funds (Levine 1997). Equally, capital markets allow investors more scope to diversify risks, and thus to support more investment and, in turn, growth.

A stable, multipillared, well-regulated financial system is a more effective and sustainable way—rather than the ad hoc targeting of specific projects or industries—of providing the right incentives for financial institutions to support economically viable (including innovative) projects rather than nonviable ones. In the international experience, such a system has often been associated with innovation, productivity gains, and sustained development. The policy implications of all of this are that, if the financial sector is to support innovation and growth, it is important for policy makers to avoid financial repression, strengthen banking sectors, develop capital markets, and build strong supporting institutions.[4] The liberalization of repressed financial markets needs to be accompanied by the establishment of adequate supporting institutions that are integral to a strong financial system. Such institutions include appropriate prudential regulation and supervision systems for banks and financial markets, strong creditor rights, contract enforcement, and good accounting practices.

China's Financial Sector Policies and Innovation

With regard to China's experience, a striking fact is that China's investment rate is very high in comparison with other countries (figure 3.2). China's investment as a percentage of GDP averaged around 40 percent during 2000–05, which was much higher than in comparator countries

Figure 3.2 *Investment as a Share of GDP, Selected Countries, 2000–05*

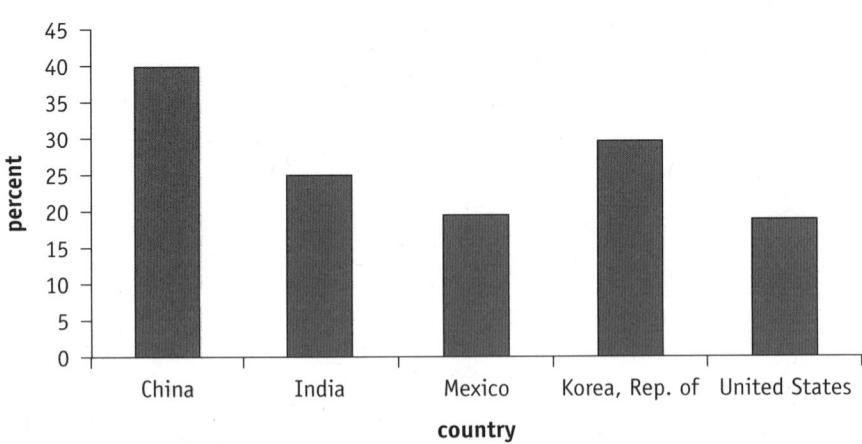

Source: IMF World Economic Outlook database.

like India and the Republic of Korea and more than double the levels in Mexico and the United States.

Investment, Overinvestment, and Efficiency

China's investment is generally considered to be at a higher level than is desirable or sustainable, and many domestic and international observers argue that, in order for growth to be sustainable, it needs to be rebalanced away from an excessive reliance on investment (and exports) and toward consumption (Aziz and Dunaway 2007; IMF 2005a, 2006a; People's Bank of China [PBC] 2007; Prasad 2007). Growth-accounting analyses suggest that China's growth in recent decades has been driven less by productivity gains and more by investment, which is susceptible to diminishing returns over time (OECD 2005).[5] A key concern is that the low cost of capital (the real interest rate is at less than half the level of the real GDP growth rate), together with the low costs of other inputs such as energy and land prices and of externalities such as pollution, has contributed to excessive levels of investment (Aziz and Dunaway 2007).[6] High levels of investment raise the risk of excess capacity in some sectors, contributing to price declines that in turn could lead to increases in nonperforming loans and associated economic and financial disruptions. Moreover, the low cost of capital has shifted production toward capital-intensive methods that entail relatively little job creation.

While investment helps to drive economic growth, a relevant consideration in all countries is whether investment is efficient. That is, if investment were allocated differently, could the same or higher rates of growth be attained with the same level of investment? Or, put another way, can the same rates of economic growth be attained with lower levels of investment?

In China, the efficiency of investment is indeed a concern. A well-functioning financial system should ensure that resources flow to the most productive uses. However, with distorted prices and a legacy of government control of financial institutions, it is questionable in China whether finance has in fact flowed to the most innovative and dynamic projects. Private enterprises, which in international experience tend to be more dynamic and innovative than state enterprises, are estimated to account for over one-half of China's GDP, but they receive only one-quarter of bank credit.[7] By contrast, state-owned enterprises produce only one-fifth of GDP, but they get two-thirds of bank credit. In other words, the less productive category of enterprises gets a higher proportion of credit than the more productive category. According to Wei and Dollar (2007), the aggregate cost of the financial misallocation is relatively large: a more efficient allocation, which entails equalizing the marginal revenue product of capital across ownership structures, would involve a transfer of two-thirds of the capital currently used by state-owned enterprises (SOEs) to the private sector and this would raise GDP by 5 percent.

Limited Access of Private Enterprises to Bank Finance

Private enterprises, as well as small and medium enterprises (SMEs), many of which tend to be private, have relatively little access to bank finance, as reflected in their low share in overall bank credit. Because banks dominate the financial system, this translates into private enterprises' low levels of access to financial sector credit overall.[8] Banks lend predominantly to large SOEs for a variety of reasons.[9] They may have concerns about their ability to accurately assess and price risk, as the clustering of loan rates around the base lending rate suggests, and thus stay away from smaller enterprises.[10] These concerns could be exacerbated by ambiguities over property rights and creditors' rights in the event of bankruptcy that could be perceived as being murkier for private than for state firms. Moreover, there could be an informal incentive structure whereby lending to SOEs is "safer" than lending to private enterprises, in the sense that if loans to the former were to go bad, then the penalties for the loan originator may be less severe.[11] The result of all this is that with large SOEs dominating banks' loan books and a limited ability of banks to price risk, there is little incentive for banks to seek out small and potentially risky borrowers.

The problem is exacerbated, as in several other countries, by information asymmetries, high transactions costs, and a lack of collateral, all of which increase SMEs' credit risk. Loan guarantees or collateral or both have become a prerequisite for most lending to SMEs in China (OECD 2005). Local governments have tried to facilitate arrangements between financial institutions and SMEs, but the experience has been uneven across provinces. Provincial governments have also established credit guarantee companies that guarantee firms' debts and in many cases perform a range of other activities. However, the performance of the credit guarantee companies has been mixed, and many are believed to be loss-making. The basic idea for these companies, first started under a pilot program in 1998, was to provide financial support for SMEs that the banks deemed to be unduly risky. The liberalization of lending rates (October 2004) may have weakened this rationale in principle; however, the fact that actual lending rates remain closely anchored to the base lending rate means that SMEs are still effectively priced out of bank credit.

Limited Access of Private Enterprises to Capital Market Finance

Private enterprises also do not have much recourse to financing from capital markets, which are still at an early stage of development. Stock market capitalization and corporate debt are much smaller than in several other countries. In 2006, the stock market surged following the end of a moratorium on initial public offerings (IPOs) and the lifting (starting in 2005) of restrictions on the tradability of shares held by SOEs. But China's stock market capitalization was still less than 50 percent of GDP, which was much smaller than in industrial countries like the United States and Japan as well as in more comparable economies like Brazil, India, and Korea (figure 3.3). Moreover, tradable shares were estimated to comprise only about a third of the market capitalization. Also, the share of equity in total funds raised, which is often a better measure than market capitalization of the role of equity finance, was only 7 percent (or about 1 percent of GDP;

Figure 3.3 *Stock Market Capitalization as a Share of GDP, 2005*

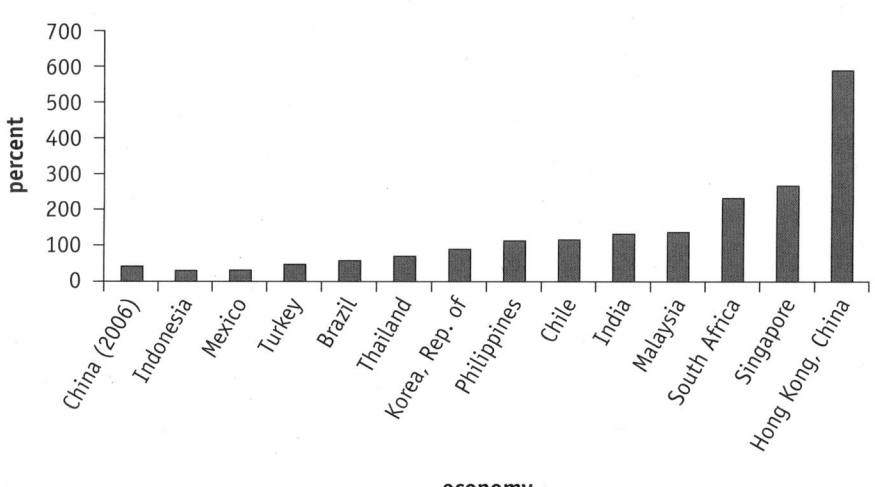

Sources: IMF 2006b; National Bureau of Statistics 2007.

78

Figure 3.4 *Source of Outside Funds as a Share of Total Funds Raised, Nonfinancial Corporations, 2005*

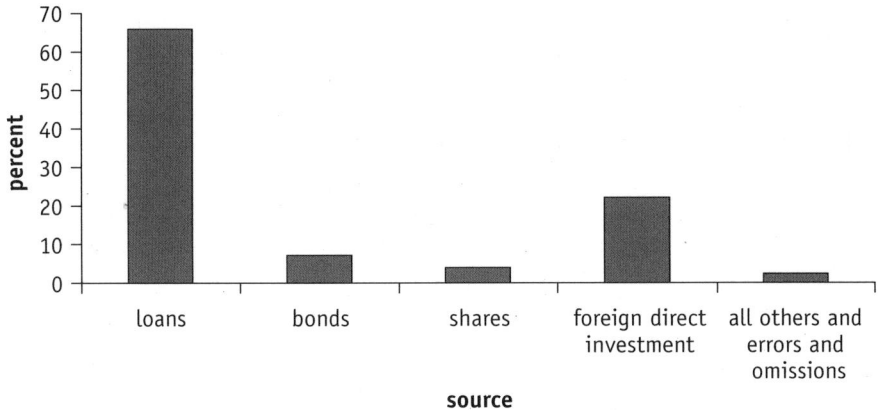

Source: National Bureau of Statistics 2007.

see figure 3.4). Corporate bonds issued in China have been equivalent to 1–1.5 percent of GDP annually in recent years. And as of mid-2006, the stock of corporate debt, at just below 15 percent of GDP, was less than in comparable Asian countries (figure 3.5).[12]

Venture capital firms, which have sometimes been effective in providing equity financing to innovative companies—including, famously, U.S. information-technology start-ups during the 1990s—have played only a small role in China. In China, they have tended to be government controlled and have often lacked technical expertise and experience.[13] In addition, until June 2007, China's company law did not allow for limited liability partnerships, which are the main forms of organization for venture capital in many countries.[14] Also, venture capital firms generally need a clear exit route to recoup their investments, such as the NASDAQ over-the-counter facility that is provided in the United States. In May 2004, a special facility for listing smaller companies was established in the Shenzhen stock exchange, but it was subject to similar restrictions as the main exchanges (including the listing requirements and nontradability of shares). In 2005, the approval requirements for foreign venture capital (and private equity) investment in domestic private firms were relaxed (SAFE 2005). Partly as a result of these changes, annual venture capital investment (foreign and

Figure 3.5 *Corporate Bonds in Selected Countries: Outstanding Stock, 2006*

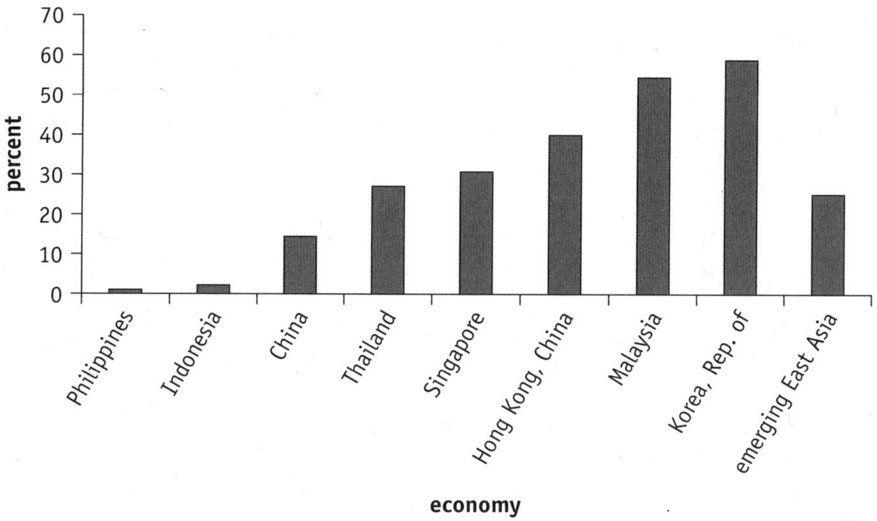

Source: Asian Development Bank 2006.

domestic) has grown from US$0.5 billion in 2004 to US$1.8 billion in 2006 (Zero2IPO Group 2006).

All in all, the capital markets play a limited role in intermediating finance to firms in China. To the extent that firms get outside finance, they get it from banks. And banks lend mainly to the large SOEs, not to smaller or private enterprises. As of March 2007, firms financed over half of urban fixed-asset investment with "self-raised funds," comprising mainly retained earnings and a very small amount of equity (figure 3.6). They financed about one-fifth of investment with bank loans. Firms financed less than 10 percent of investment through foreign capital and allocations from the state budget, and a negligible amount through corporate bonds. As in many developing countries, an informal financial sector operates in China to serve firms and individuals that have difficulty accessing the formal financial sector, particularly SMEs and farmers. The exact size of the informal sector is hard to pin down, as the sector falls outside the formal regulatory framework, but according to one survey it stood at 5–6 percent of GDP in 2003 (Central Finance University of China 2005).

Figure 3.6 *Sources of Financing as a Share of Total Urban Fixed-Asset Investment, 2007*

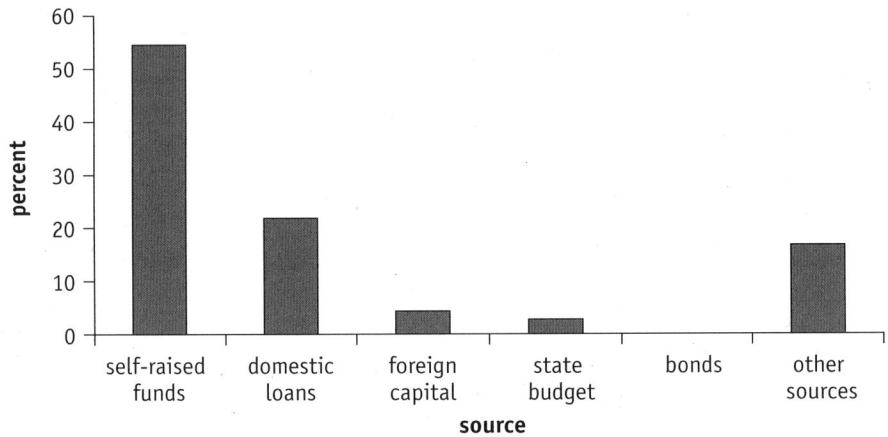

Source: National Bureau of Statistics 2007.
Note: Bank loans are included as domestic loans and as a small part of other sources.

Reform Implications and Policy Agenda

Against this background, financial sector reforms in China could have significant implications for efficiency and productivity in the real economy. A reformed banking system that is adequately capitalized, commercially oriented, and well supervised and regulated would finance a wider range of firms, beyond just SOEs, and a more efficient set of economic activities. Expanded capital markets would help reduce firms' need for internal financing, and by giving larger enterprises an alternative way to raise funds, they could increase the scope for banks to lend to small enterprises, which often are sources of innovation. Reforms could help widen the range of credit instruments for households and increase financing for small businesses, because they often tend to be household run (Tran 2006).

The Chinese authorities are well aware of the challenges in the financial sector and are moving ahead with wide-ranging reforms in a number of key areas, particularly in the banking system. Three of the four state-owned commercial banks have been restructured and recapitalized, and have held successful IPOs. The transformation of nontradable into tradable shares is contributing to a revival of the equity market, and some early steps have been taken to develop fixed-income markets. However, much remains to be done to further strengthen the banking system, particularly its commercial orientation and corporate governance, and to fill in other gaps in the financial sector—such as development of the corporate bond market and of the rural financial system—in order for the financial system to more fully support the overall economy, including smaller firms.

The reform objectives were clarified most recently at the National Financial Work Conference in January 2007, which comprehensively identified the challenges that remain in the financial sector and set a clear agenda for reforms in the period ahead. Key elements of the agenda include the need to (1) continue with bank reforms, notably by further strengthening corporate governance in the state-owned commercial banks, reforming the Agricultural Bank of China, and commercializing the policy banks starting with the China Development Bank; (2) develop the corporate bond market; (3) strengthen the environment for SMEs; and (4) develop the rural financial system.[15]

Banking Sector Reforms   Bank reforms have been at the forefront of the strategy for financial sector reform, and much has been accomplished in recent years (Zhou 2005). In 2005, financial restructuring was completed for three of the four state commercial banks, which reduced their nonperforming loans; the banks were recapitalized and subsequently attracted strategic foreign investors and conducted successful IPOs.[16] The rural credit cooperatives are being reformed, and other elements of the rural financial system are being developed. Banking supervision is being strengthened under the lead of the China Banking Regulatory Commission, which was formed in 2003. The commission has supervisory and regulatory authority over commercial and policy banks, as well as over nonbank financial institutions and the rural financial sector.

Several challenges remain, however. It is important for regulators to ensure the enforcement of prudential and supervisory rules and the sound monitoring and supervision of the flow of new nonperforming loans to prevent the recurrence of the situation that prevailed before the restructuring of the large state commercial banks. It is also important to ensure that regulators and banks have the ability to monitor and manage credit and foreign exchange risk. Improving banks' corporate governance and commercial orientation remains an ongoing challenge. The Agricultural Bank of China still has to be reformed, and a delay in reforming it risks exacerbating a moral hazard that could undermine the reforms of the other major banks. Although the intention to reform the agricultural bank has been announced, a substantive reform plan with restructuring and recapitalization measures, along with a timetable for implementing the plan, are still awaited.

Equity Market Development   In recent years the authorities have taken several measures to develop the equity market. The old system under which new listings were rationed has been replaced since 2000 with a procedure under which companies have to meet objective economic criteria set by the China Securities Regulatory Commission to become eligible to list. However, final approval for each listing is still required from the government, and the system is not "disclosure based." As a result, the approval process for IPOs can be protracted, as can the period between approval and actual listing. The tradability of previously nontradable shares of the state in listed companies since 2005 has contributed to the surge in activity and prices during the subsequent period.[17] Disclosure requirements are being streamlined, which will make it easier for small firms to list in the market. However, there is still some way to go in market development, as the experience with a growing market is still very recent (before 2006 the stock market had been moribund since the slump of 2001), tradable shares still account for a relatively small proportion of market capitalization, and equity represents only a small part of firms' total financing in China compared with other countries. In addition, investor knowledge and familiarity with the market are still at an early stage.[18]

Bond Market Development   More progress has been made in developing the market for shorter-term securities than for longer-term securities. The People's Bank of China (PBC) opened a short-term corporate bill market and established an interbank market for asset-backed securities in 2005. Although the latter has yet to take off, the corporate bill market has developed significantly. However, the (longer-term) corporate bond market remains shallow and illiquid. Some lessons from other countries might be instructive (see box 3.1).

In China, a pilot program was launched in August 2007 to address several of the key impediments that have held up corporate bond market development. The program moves the bond issuance procedure for listed companies much closer toward a disclosure-based system that is followed in many countries and streamlines the regulatory burden. Indeed, a key impediment to corporate bond market development in China has been the high regulatory burden. Regulatory authority has been scattered across different agencies—the National Development and Reform Commission is responsible for primary issues,

Box 3.1 *International Experience with Corporate Bond Market Development*

Corporate bond markets are relatively recent phenomena in most countries, with the United States and Germany being notable exceptions. In Europe, corporate bond markets have generally grown rapidly only after the introduction of the euro. In Japan, the market has developed mainly after the deregulation that started in the mid-1980s, which entailed a relaxation of market eligibility standards, establishment of ratings agencies, and start of bond futures trading, followed by the "Big Bang" financial reform in the mid-1990s. In emerging markets, the macroeconomic and financial crises in the 1990s led to greater efforts to develop local bond markets as an alternative source of financing for the corporate sector and as a way to reduce currency and maturity mismatches on corporate balance sheets. As a result, corporate bond markets grew substantially in several emerging economies, particularly in Asia and Latin America. The Republic of Korea and Malaysia now have two of the largest such markets in the world relative to GDP, although in most emerging economies, such markets are still small.

In emerging economies, both cyclical and structural factors have played a role in corporate bond market development. In the aftermath of the financial crises of the 1990s, corporations faced a need to restructure and to find alternative sources of financing. Lower inflation and interest rates created a conducive environment for corporate entities to refinance expensive external debt with cheaper local debt. This was supported by strong growth in assets under management of local institutional investors, such as pension funds, insurers, and asset management companies. In addition, many authorities implemented reforms specifically aimed at the development of corporate bond markets, such as establishing ratings agencies, easing market eligibility standards, establishing benchmark yield curves, and strengthening the market infrastructure through the use of trading platforms, clearing and settlement systems, the regulatory environment, and other methods.

In general, the establishment of well-functioning corporate bond markets has entailed overcoming a series of challenges. Macroeconomic stability has been a key precondition, as large fiscal deficits tend to crowd out corporate issues and to raise yields. A low-inflation environment helps to keep yields stable and low. A large and diversified institutional investor base has been generally helpful to ensure a steady demand for corporate issues. In turn, a set of diverse issuers with adequate size, credit quality, and transparency is needed to meet the demand from institutional investors. A regulatory framework that ensures investor protection and market integrity—including bankruptcy laws that clearly define creditors' rights and debtors' responsibilities as well as good practices in corporate governance and disclosure—is important for the effective functioning of securities markets. And markets need to be supported by an adequate microstructure for primary and secondary issues.

Source: IMF 2005b.

and the China Securities Regulatory Commission is responsible for the trading of listed bonds—and the approval process has been lengthy. Primary issuance has been subject to an annual quota, which has, however, been eliminated for listed companies under the pilot program.

The use of a merit-based system for issuance—whereby the government approved every issue—rather than a disclosure-based system made it hard for firms, and particularly smaller firms, to issue bonds. As a result, virtually all corporate bond issuers have been SOEs. Regulatory policies focused on determining the particular characteristics of bonds (such as maturities and yields) rather than on ensuring that issuers provide adequate disclosure of their financial conditions. Regulatory control over bonds' features were somewhat liberalized over time, but key features remained tightly regulated before 2007. Initial bond yields, for example, were subject to a regulatory cap and could not be more than 40 percent higher than the bank rate for deposits of similar maturity. (This requirement was also eliminated under the pilot reform program.)

Small and Medium Enterprises   The financial system thus provides only limited support for small and medium enterprises, and for private enterprises. The incentives for banks and the regulatory framework in the capital markets have tended to favor large SOEs over other firms. In this context, the government's moves to develop a well-diversified financial system, with sound banks and deep capital markets, and to undertake related reforms in a number of areas are very much in the right direction and should help to create a more level playing field across different types of enterprises.

The passage of relevant legislation in 2007 and continuing measures to improve the availability and transparency of debtor information may help to improve the environment for lending to SMEs. The new Property Law addresses some of the property rights issues that have constrained lending to SMEs, and the Bankruptcy Law gives secured creditors the senior-most claim in the event of bankruptcy. The PBC has created a national credit database that should reduce banks' loan-assessment costs and allow them to price risk better. It established a central credit database in 1997 that by 2002 covered corporate borrowers in most of the country. In 2006, it made operational a nationwide credit registry for all loans to individuals, which was then merged with the corporate database into a unified system.

The establishment in the Shenzhen stock exchange of the Small and Medium Enterprise Board in 2004 was a potentially important step toward giving small companies greater access to equity finance and exposing them to more market discipline. The SME Board has been subject to the same restrictions as the main exchanges, but it is expected in the future to have separate simplified and streamlined listing requirements as well as its own market infrastructure, including a separate trading system and index. Separately, at the 2007 National Financial Work Conference, the government spoke in favor of developing a multitiered capital market, including an over-the-counter market that could provide a way for unlisted companies to transfer their ownership rights.

Conclusion

The general international experience suggests that financial development positively influences economic growth and development. It does so mainly by contributing to greater productivity, in part by supporting and encouraging innovation, including on the part of smaller-scale enterprises. In China, savings are very high as a proportion of income, but the efficiency with which those savings are intermediated by the financial system is relatively low. The system appears to channel credit disproportionately toward large SOEs, rather than toward smaller and private enterprises, because of both the incentives in the banking sector and the regulatory setup of the capital markets.

Some of the key imperatives going forward would appear to include the following:

- Reform of the Agricultural Bank of China, and continued improvements in the operations and supervision of the other banks to ensure that the recent strengthening in their balance sheets is sustained
- An expansion in the volume of tradable shares in the equity market, and a streamlining of the process for issuing equities in order to allow participation by a wider group of companies
- A continued easing of the regulatory burden for issuing corporate bonds, including a move from the merit-based to a disclosure-based system.

China has made steady progress in financial sector reforms in recent years. A continued implementation of the reform agenda in the period ahead, and the attainment of the objectives laid out at the 2007 National Financial Work Conference, should help to reverse the current bias toward large state-owned enterprises, create a more level playing field for all enterprises, and help to improve the environment for innovation and sustained growth.

Notes

This chapter is based on the author's presentation at the 2006 Asia-Pacific Finance and Development Center conference in Shanghai. He would like to thank Steven Dunaway, Qimiao Fan, Li Kouqing, Richard Podpiera, and conference participants for helpful comments and discussions. The views reported in this chapter are those of the author alone and should not be attributed to the IMF or its executive board.

1. The term *financial development* as used in this chapter refers to the development of a banking sector that is characterized by competition, commercial orientation with strong internal controls and governance, and sound supervision and regulation, and of deep and liquid capital markets, particularly bond and equity markets.

2. Figure 3.1 shows bank credit as a percent of GDP, which is a commonly used measure of the level of financial development. However, bank credit can sometimes simply reflect a financial system that is overly dependent on banks rather than capital markets. Nonetheless, a number of alternative measures (such as capital market sophistication) also illustrate the positive relationship between financial and economic development.

3. See IMF (2004) for a review of the literature.

4. *Financial repression* refers here to policies such as directed lending, heavily controlled interest rates, and government interference in financial institutions' decisions, all of which prevent such institutions from operating on a commercial basis.

5. Indeed, some estimates suggest that the marginal product of capital declined by about 3 percentage points during 1993–2004 (IMF 2005a). Although other estimates suggest a slight increase in the marginal product of capital since the early 1990s, they do not dispute the conclusion that the level of investment is high relative to what is considered sustainable (He, Zhang, and Sheck 2006).

6. Prasad (2007) also provides a comprehensive discussion of the challenges associated with rebalancing China's economy.

7. See Farrell (2006) and OECD (2005), which also discusses the alternative definitions of "private" enterprise in China. This is more complicated than one might think, because while relatively few enterprises are formally registered as strictly private, many more are in fact not directly under state control (for example, joint ventures, shareholding corporations).

8. In 2006, China's bank credit as a percentage of GDP stood at nearly 140 percent, about 40 percentage points higher than the average for Asian countries. Stock market capitalization was about a third this size, and the stock of corporate bonds was only a 10th this size.

9. The joint stock banks and city commercial banks, however, lend proportionately more to SMEs than do the state-owned commercial banks.

10. Podpiera (2006) found that the pricing of credit risk by banks is relatively undifferentiated across borrowers.

11. See Prasad (2007), who also discussed the overall macroeconomic situation and challenges in China.

12. This figure includes corporate bonds issued by both financial and nonfinancial institutions. The stock of bonds issued by nonfinancial institutions is equivalent to less than 1 percent of GDP.

13. See (OECD 2005). Also, the transparency of venture capital (and private equity investment) funds needs to be adequate for supervisory and regulatory purposes before such funds can develop substantively.

14. Effective June 1, 2007, the new Partnership Enterprise Law (2007) allows Chinese firms to invest in venture capital firms as limited partners.

15. Rural financial reform is an important part of the reform agenda and is needed to deepen financial development across the country, but it is outside the scope of this chapter.

16. After being recapitalized, the banks were restructured into shareholding companies, and then they conducted IPOs. China Construction Bank (CCB) listed its shares in Hong Kong, China, in 2005, and the Bank of China (BOC) and the Industrial and Commercial Bank of China (ICBC) each listed their shares jointly in Hong Kong, China, and Shanghai in 2006.

17. Apart from the reforms, strong corporate earnings and market sentiment have also played a role, as has the fact that the newly tradable shares have come into the market only gradually, in part because they were subject to lock-up periods.

18. A recent market survey, for example, found that more than half of investors had "little or no knowledge" of how the stock market operates (see *China Daily* 2007).

References

Asian Development Bank. 2006. *Asia Bond Monitor*. November. Manila: Asian Development Bank. http://asianbondsonline.adb.org/documents/abm_nov2006.

Aziz, J., and S. Dunaway. 2007. "China's Rebalancing Act." *Finance and Development* 44 (3).

Beck, T., R. Levine, and N. Loayza. 2000. "Finance and the Sources of Growth." *Journal of Financial Economics* 58 (1–2): 261–300.

Central Finance University of China. 2005. "Nationwide Survey on Informal Finance." Unpublished, Central Finance University, Beijing.

China Daily. 2007. "Traders Be Warned." May 31. http://www.chinadaily.com.cn/china/2007-05/31/content_883867.htm.

Farrell, D. 2006. "China's Financials Need an Overhaul." *BusinessWeek,* August 9. http://www.mckinsey.com/mgi/MGInews/businessweek/chinafinancial.asp.

Greenspan, A. 2002. "Regulation, Innovation, and Wealth Creation." Remarks presented at the meeting of the Society of Business Economists, London, September 25.

He, D., W. Zhang, and J. Sheck. 2006. "How Efficient Has China's Investment Been? Empirical Evidence from National and Provincial Data." Research Memorandum 19/2006, Hong Kong Monetary Authority, Hong Kong, China.

IMF (International Monetary Fund). 2004. "Are Credit Booms in Emerging Market Economies a Concern?" In *World Economic Outlook* (April).Washington, DC: International Monetary Fund.

———. 2005a. "People's Republic of China: 2005. Article IV Consultation." Country Report No. 05/411, IMF, Washington, DC.

———. 2005b. "Development of Corporate Bond Markets in Emerging Market Countries." In *Global Financial Stability Report* (September). Washington, DC: IMF.

———. 2006a. "People's Republic of China: 2006." Article IV Consultation, Country Report No. 06/394, IMF, Washington, DC.

———. 2006b. "Asian Equity Markets: Growth, Opportunities, and Challenges." *Asia and Pacific Regional Economic Outlook,* September. Washington, DC: International Monetary Fund.

King, R., and R. Levine. 1993. "Finance and Growth: Schumpeter Might Be Right." *Quarterly Journal of Economics* 108 (3): 717–38.

Levine, R. 1997. "Financial Development and Growth: Views and Agenda." *Journal of Economic Literature* 35 (2): 688–726.

National Bureau of Statistics. 2007. *China Statistical Yearbook 2006.* Beijing: National Bureau of Statistics.

OECD (Organisation of Economic Co-operation and Development). 2005. *OECD Economic Surveys: China.* vol. 2005/13. Paris: OECD Publishing.

PBC (People's Bank of China). 2007. *China Monetary Policy Report: Quarter 1, 2007.* Beijing: People's Bank of China.

Podpiera, R. 2006. "Progress in China's Banking Sector Reform: Has Bank Behavior Changed?" IMF Working Paper 06/71, International Monetary Fund, Washington, DC.

Prasad, E. 2007. "Is the Chinese Growth Miracle Built to Last?" Draft, Cornell University, Ithaca, New York.

Rajan, R., and L. Zingales, 1998. "Financial Dependence and Growth." *American Economic Review* 88 (June): 559–86.

SAFE (State Administration of Foreign Exchange). 2005. *Relevant Issues Concerning Foreign Control on Domestic Residents' Corporate Financing and Roundtrip Investment through Offshore Special Purpose Vehicles*. SAFE Circular No. 75. Beijing: State Administration of Foreign Exchange.

Schumpeter, J. 1912/1949. *The Theory of Economic Development: An Inquiry Into Profits, Capital, Credit, Interest, and the Business Cycle*. Trans. Redvers Opie. Cambridge, MA: Harvard University Press.

Tran, H. 2006. "Financial Sector Reforms in China and India." Remarks at the "Recent Research on Hedge Funds" Conference, Northwestern University, Evanston, Illinois, August.

Wei, S-J., and D. Dollar. 2007. "Das (Wasted) Kapital: Firm Ownership and Investment Efficiency in China." IMF Working Paper No. 07/9, International Monetary Fund, Washington, DC.

Zero2IPO Group. 2007. *China VC, PE, M&A, IPO Report 2006*. Beijing: Zero2ipo Group. http://www.zero2ipo.com.hk/Venture_Knowledge_Center/Z-Report.asp.

Zhou, X. 2005. "Speeding up Financial Reform to Support Economic Development of the West Region in China." Speech at the "High-Level Forum on Economic Development and Financial Services," Chongqing, China, November 15.

4

The Innovation and Development of China's Bond Market

Huaipeng Mu

Building an innovative country is of strategic importance to China's economic development. The concept of innovation covers both scientific and technological innovation and that of social institutions. As a crucial sector in China's national economy, it is important for the financial sector to innovate itself and to support innovation in other sectors. Since China's reforms and opening up, the financial sector has achieved rapid development. The product innovation and financial market's institutional innovation played a very important role in the process of economic system reform and in the continuous deepening of financial reform and opening up. This chapter primarily is an introduction to product innovation and institution building in China's bond market over the past two years, as well as to the developments in the bond market.

Product Innovation and Expansion

Different financial products have different capabilities in raising capital and diversifying risks. One of the benchmarks of maturity in a financial market is the market's ability to offer diversified financial instruments to meet the needs of diverse fund-raising entities. Since 2004, the People's

Bank of China (PBC) has made product innovation and increased financing instruments a focus of the developing financial market, made active explorations in bond market innovation, and achieved significant progress.

Innovation in Bond Products

The first product is the securities companies' short-term financing bills. In October 2004, the PBC issued the Regulations on Short-Term Financing Bills Issued by Securities Companies, which permits securities companies to issue short-term financing bills to institutional investors. On April 12, 2005, Guotai Junan Securities Company issued short-term financing bills for 600 million yuan (Y). In 2005, five securities companies issued short-term financing bills of Y 2.9 billion, which have been redeemed.

The second product is the subordinated bonds issued by commercial banks. To improve asset quality and capital adequacy levels of commercial banks, as well as to advance the shareholding reform process of the state-owned commercial banks, the PBC issued the Regulations on the Issuance of Subordinated Bonds by Commercial Banks in June 2004. At that time, the PBC granted approval for 14 banks, including Bank of China, China Construction Bank, Industrial and Commercial Bank, Industrial Bank, and China Minsheng Banking Corporations, Ltd., to issue subordinated bonds. By the end of October 2006, the total value of subordinated bonds issued by commercials banks had reached Y 164.7 billion.

The third product is the ordinary financing bonds issued by commercial banks. To expand direct financing channels for financial institutions and provide them with the liability management tools needed, and to resolve their long-standing problems of terms mismatch between savings and loans, the PBC, after considerable public consultation, issued the Regulations on the Issuance of Financial Bonds in the National Interbank Bond Market on April 27, 2005. These regulations allow commercial banks and financing companies of corporate groups to issue ordinary bonds. By the end of December 2006, Shanghai Pudong Development Bank, China Merchant Bank, Xingye Bank (Industrial Bank), and Minsheng Bank had issued financial bonds worth Y 58 billion.

The fourth product is credit asset securitization. Asset securitization is the structured process in which assets (that is, residential mortgage, car mortgage, credit card receivables, corporate receivables, and other assets such as infrastructure and intellectual property) that can generate future cash flows are packaged, underwritten, and sold in the form of asset-backed securities.

On April 21, 2005, PBC and the China Banking Regulatory Commission jointly issued the Regulations on Pilot Credit Asset-Securitization. China Development Bank was the first Chinese bank used for trial issuance of mortgage-backed securities and residential mortgage-backed securities. By the end of December 2007, China Development Bank and China Construction Bank had issued Y 10 billion worth of credit asset–backed securities and Y 3.02 billion of residential mortgage–backed securities, respectively. This was followed by two issuances, a total of Y 5.85 billion, of nonperformance-loan-backed securities by two asset management companies, China Oriental and Xinda.

The fifth product is hybrid capital bonds. In September and December 2006, Industrial Bank and Minsheng Bank issued Y 4 billion and Y 4.3 billion of hybrid capital bonds, respectively. The issuance of subordinated bonds and hybrid capital bonds has played a very important role in advancing the shareholding reform of the state-owned commercial banks and in improving the capital adequacy level of the state-owned commercial banks. These hybrid bonds also have become an important channel through which commercial banks can supplement their subsidiary capital.

The sixth product is the nonfinancial corporations' short-term financing bills. In May 2005, the PBC issued the Regulations on Short-Term Financing Bills and other related supporting documents in May 2005. Soon after, qualified enterprises started to issue short-term financing bills to institutional investors in the interbank market, aiming to provide direct financing for nonfinancial corporations. On May 26, 2007, six enterprises, including Huaneng International, issued the first batch of short-term corporate financing bills, with a total value of Y10.9 billion. By the end of December 2006, 210 enterprises had issued short-term financing bills close to Y 433.65 billion. The issuance and circulation of short-term financing bills are a major breakthrough in expanding financing channels, improving China's finance structure, and promoting corporate bond market development.

Innovation in Trading Instruments and Financial Derivatives

As the first innovation, outright repos (repurchase agreements) have been introduced for the purpose of increasing bond market liquidity. On April 12, 2004, the PBC enacted the Regulations on Bond Outright Repos in the National Interbank Bond Market to lay down the structure of the outright repo. Outright repo trading has not only increased the liquidity of the bond market, but has also paved the way for the introduction of other derivatives. Turnover of outright repos in the interbank bond market totaled Y 347.66 billion by the end of 2005 and Y 291.88 billion by the end of 2006.

Second, bond forward transactions were recently introduced. A bond forward transaction contract is a legal agreement between the two transaction parties who agree to buy or sell bond assets at a fixed price at a future time. The transaction is based on the different expectations of both parties regarding future exchange rates and market conditions and can help investors manage interest-rate risk. In addition, the forward trading can provide key information for the central bank in its conduct of monetary policy and can play an important role in promoting the future development of the bond market, in particular, and the financial market, generally. In 2006, the turnover of bond forward transactions totaled Y 65.85 billion. Consequently, the Regulations on Bond Forward Transitions in the National Interbank Bond Market was enacted on May 17, 2005, and came into effect on June 15, 2005.

Third, the Y interest rate swap in the Notice of the People's Bank of China on the Launch of Y Interest Rate Swaps on Pilot Basis (January 24, 2006) is defined as two parties exchanging their cash flows that arise from the agreed-upon principal amount in a specific period of time. The cash flows of one party are calculated at a floating rate, and the cash flows of the other are calculated at the fixed rate. The PBC enacted the notice to diversify risk management tools for investors in the interbank bond market, to regulate and direct Y interest rate swaps, and to accelerate market-oriented interest rate reform. By the end of February 2007, the amount of interest rate swaps had reached Y 27.01 billion.

In the fourth innovation, in November 2006, the PBC enacted the Provisional Regulations on Management of the National Inter-Bank Market Bond Lending Business and thereby officially introduced bond lending business into the interbank market. This innovation is conducive to enhancing market liquidity, reducing settlement failure, and maintaining market stability. It also provides investors with a new profit-generating model and risk-aversion instruments, therefore increasing market effectiveness.

Institution Building and Financial Market Development

Institutions, as a kind of norm, can provide market participants with reasonable expectations, reduce transaction costs, and raise transaction efficiency; thus they are the safeguards for the healthy development and continuous innovation of financial markets. The PBC has all along attached great importance to the institution building of financial markets.

Since 2004 the PBC has intensified efforts in bond product innovation and also accelerated institution building in the field of interbank bond markets. The following aspects have been put in place and strengthened.

First, the management mechanism for issuance of short-term financing bills has been established. This new institutional arrangement is mainly manifested in the market-oriented reforms in the fields of market access, issuance modality, intermediary services, operational processes, and risk prevention. On market access, the new filing management system means that it is the bond-issuing enterprises, rather than the government authorities, that decide whether there is any need for debt issuance. There are no discriminatory provisions against various categories of enterprises. In terms of issuance modalities, an enterprise first looks for a major underwriter ready to conduct debt underwriting, which is in turn responsible for making arrangements for the issuance. The targeted investors are institutional investors on the interbank bond market, and the price is determined by the bid/ask and book-entry methods. Regarding risk prevention, information disclosure and a rating system have been introduced instead of bank guarantees. This practice has resulted in strengthened market-disciplining mechanisms and roles played by intermediaries. Practice has demonstrated that in the process of developing the corporate bond market, it is feasible to adopt a market-oriented filing management system that is conducive to the market development.

Second, more stringent information disclosure requirements have been imposed. This has been an important aspect of the institution building of markets over the past few years. More stringent information disclosure is intended to address the problem of information asymmetry between investors and fund-raisers and to subject the issuers to the joint supervision of market participants. The requirements on more stringent information disclosure have been imposed since the establishment of the interbank bond market in 1997. In October 2002, a uniform set of norms of information disclosure was imposed on a trial basis on some securities firms in the national interbank market. In 2003, these norms were applied to all securities firms regarding information disclosure.

Ever since 2004, in all the newly enacted bond regulations, there have been stringent information disclosure requirements, including the PBC's "Note Regarding Information Disclosure on Short-Term Financing Bills" and other specific regulations regarding information disclosure. Among these regulations and notices are provisions concerning issuance-information disclosure, continuous information disclosure, the temporary notice on major events, the disproportionate shareholding notice, and the default fact notice.

Regarding implementation of information disclosure, specific regulations address the platform, procedures, exemption, and violation penalties of information disclosure. Practice has demonstrated that, compared with the one-to-one information disclosure and unilateral restraining mechanism existing between a bank and its client in the course of bank lending, it is more effective to adopt a one-to-multiple information disclosure mechanism where fund-raisers make information public to all market participants.

Third, the bond rating system has been introduced. Prior to 2004, only treasury bonds and policy bank bonds were issued and traded on the interbank bond market and there was no urgent need for bond rating. Since 2004, after securities firms were permitted to issue short-term financing bills and commercial banks were allowed to issue subordinated bonds, more and more financial products with credit risk were introduced into the interbank bond market, which makes the establishment of a bond rating system increasingly important. The role of rating is to rate the potential risks contained in the bond products so that investors can make better decisions about the desirability of investment and then determine the prices of bonds. Ratings are required in all the regulations on financial bonds, enterprises' short-term financing bills, asset-backed securities, and renminbi-denominated bonds issued by the international development institutions. In particular, credit rating has become an important aspect in the issuance and management of the enterprises' short-term financing bills.

The Development of China's Bond Market

With the accelerated pace of product innovation and institution building, the bond market has achieved rapid developments, which can be demonstrated in the following aspects of the market.

First, the market size has continued to expand. The volume of bond issuance and trading has increased by a big margin. In addition, the market's resource-allocation function has been brought into full play. In 2006, about Y 5 trillion worth of bonds was issued in China's bond market, an increase of 33.2 percent over the same period in the previous year. By the end of 2006, the amount of custody on China's bond market had reached Y 9.2 trillion, increasing by 27.5 percent compared with the previous year. The vast majority of bonds had been issued on the interbank bond market, with the bonds in the custody of the interbank bond market representing 96.1 percent of the total amount of bonds issued in China. Bond trading has been active and market liquidity has continued to increase. In 2006, the turnover of outright repo in the bond market totaled Y 28.2 trillion, a rise of 54.6 percent over the same period the year before. Out of this amount, turnover of outright repo in the interbank bond market totaled Y 26.6 trillion, which is 94.4 percent of the entire turnover. On China's bond market, the turnover of coupon bonds totaled Y 10.4 trillion, an increase of 65.4 percent from the previous year, and of this total, the coupon bond turnover in the interbank market totaled Y 10.3 trillion, which accounts for 98.4 percent of the total turnover.

The second aspect is that the body of bond issuers and investors has been greatly diversified. As the pace of market liberalization has accelerated, the number and diversity of issuers and investors have burgeoned. The bond issuer in the interbank market is an example: in recent years, in addition to the original issuers such as the Ministry of Finance and policy banks, other institutions have already started to issue bonds on the interbank bond market, including the state-owned commercial banks, shareholding commercial banks, city commercial banks, securities companies, nonfinancial institutions, and international development institutions. At the same time, the investors are also burgeoning, with institutional investors performing actively. In addition to the financial institutions, some nonfinancial institutions, such as fund management companies, accounting firms, and insurance companies, and some enterprises have rapidly increased their participation in and their influence on the bond market. By the end of 2006, the number of market players in the interbank market had increased from 5,508 at the end of 2005 to 6,279, an increase of 14 percent.

In the third aspect of development, the market has further deepened its functions and gradually expanded its influence. The regulatory function of the bond market has gradually become significant, which allows it to become one of the important platforms for the Chinese government to conduct macroeconomic regulations and implement a prudent monetary and fiscal policy. In 2006, the Ministry of Finance issued Treasury bonds of Y 880 billion, which provided a solid basis for implementing prudent fiscal policies. The PBC issued central bank bills of Y 3.65 billion on the bond market, which effectively adjusted the liquidity in the banking system. By 2006, short-term financing bills of Y 292 billion had been issued, counting for 7.8 percent of total corporate financing in the same year and largely increasing the scale of corporate direct financing.

Fourth, the bond market's function as a price signal has become increasingly significant, and the yield curve has gradually come into shape. Since the interest rate formation mechanism on the bond market has been fully market based, changes in both the bond price index and the yield curve can effectively reflect the status of supply and demand and liquidity on the bond market. Meanwhile, the formation of the yield curve has laid the foundation for the pricing of financial assets.

The Main Experiences of China's Bond Market Development

The PBC has conducted scientific and coordinated planning for the development of the interbank bond market, drawing upon the advanced international experiences and the over-the-counter (OTC) bond market model. The main lessons of China's bond market development can be summarized as follows.

First, it is imperative for readers of this chapter to fully recognize the importance of developing the bond market. Since the beginning of the reform and opening-up process, China's financial markets have been developing rapidly, and a relatively complete market system has been put in place. However, the market has been encountering unbalanced development, characterized by slow direct financing, the disproportionate ratio of direct financing to indirect financing, and the heavy reliance of companies on indirect financing. Over the past decade, China's direct corporate financing has remained at 10 percent or so. Among direct financing, the percentage of corporate bonds has been far lower than that of equity financing. In 2006, corporate debts represented only 44 percent of the total amount of equity financing in that year. On the bond market, the proportions of government bonds and financial bonds were higher than that of the corporate bonds, which account for only 6.77 percent of the entire debt issuance.

Compared with other countries, this corporate bond ratio may seem very low, yet it has not affected China's economic development. However, it is necessary to recognize that direct financing is associated with financial stability. In a system dominated by indirect financing, financial risks are excessively concentrated in the banking sector and thus create a hidden source of systemic financial risks. Direct financing instruments, in contrast, can be traded, and thus risks can be mitigated by market mechanisms. Even if some losses are incurred, investors will absorb them, and risks will not be magnified. In addition, under the circumstance of the modern economy, direct financing is a convenient and highly effective way of transforming the society's savings pool into long-term investment and promoting corporate capital formation and expansion. It has become an imperative choice to promote bond market development, in particular the corporate bond market, to effectively allocate financial resources, strengthen market disciplines, diversify and mitigate risks, and maintain financial stability.

Second, it is critical to promote innovation in the developing bond markets, since innovation has become an important driver for bond market development. In recent years, the depth and width of the interbank bond market have increased, market liquidity has continuously increased, the market function has deepened, and the market influence has continuously expanded. These achievements are inseparable from the PBC's emphasis on promoting financial market innovation. Innovation requires deregulation. If all the financial products are subject to approval, there will be no innovation. In a very complicated financial operational environment, deregulation will be conducive to innovation. Meanwhile, it is necessary to keep a balance between deregulation and regulation in promoting indigenous innovation. An important lesson is to raise standards; strengthen the rules of information disclosure, credit rating, and external auditing; and implement accounting rules.

It is essential to help the market accept new ideas so that innovation can prosper in a friendlier environment. Innovation is a very broad concept, which includes innovation in market products and instruments and in market institutions. In addition, innovation can be driven either by the government agencies or by the market. Along with the continued improvement in the market economic system, the market should promote future innovation of market products and instruments, while institutional innovations regarding information disclosure, credit rating, accounting, and taxation should primarily be completed by the government authorities. In this context, the relevant government agencies and market participants should all actively embrace new ideas, further streamline and adjust their respective functions and positioning, and make joint efforts to establish a long-term mechanism for innovation on the bond market.

The third lesson is that, in the development of the bond market, it is important to target institutional investors as the principal investors and to give top priority to introducing and fostering institutional investors. The fact that the bond market is dominated by institutional investors is determined by the nature of bond products. On the one hand, bonds are one kind of fixed-income instrument with clearly defined maturity dates, stable future cash flows, and relatively low returns but relatively high safety. For the small amount of funds of individual investors, the returns are not obvious. On the other hand, bonds are characterized by large varieties, complex and diverse transaction modalities, and highly innovative instruments, which require rich expertise and investment techniques possessed by institutional investors. Therefore, individual investors are in a less-advantaged position in bond market investment. Generally speaking, individual investors can share benefits of market development by pooling funds. International experiences indicate that, in the mature market system, including the United States, the United Kingdom, and Japan, the percentage of individual investors does not exceed 5 percent, and in emerging markets such as the Republic of Korea, the percentage is only 10 percent. Therefore, the predominance by institutional investors is a major characteristic and inevitable result of bond market development. In China's own development experience, after the launch of the interbank bond market in 1997, the PBC exerted great effort to introduce and foster institutional investors and promote the development of the interbank bond market, following the lessons learned in the development of China's bond market and the standard rules of international bond markets. At present, as an important component of the bond market, the interbank bond market has in fact become a market for institutional investors.

The fourth lesson is that the development of the bond market should primarily rely on the OTC market and promote the connection between the open market and the OTC market. The fact that the OTC market is the main component of the bond market is determined by the rule in the development of the bond market. The rule is that the mainstream of the bond market is institutional investors, who trade in large volumes; such large-volume trading orders are flexible and diverse, are more easily matched on the OTC market, have relatively lower trading costs, and lead to higher market efficiency. International experience has demonstrated that the vast majority of bond transactions in advanced countries are all done on the OTC market, with the open market mainly functioning as price signals and credit enhancement for issuers. The bond market in the United States, the United Kingdom, and Japan almost exclusively trade on the OTC market, and in Korea and other emerging markets the turnover on the OTC market is also over 95 percent of all the bond turnovers.

In China, the interbank bond market, which is China's OTC market, has drawn upon the best practice from developed bond markets around the world and adopted the bidding-based trading system that is suitable for institutional investors. In 2005, the turnover of coupon bonds in the OTC market accounted for 95.6 percent of total transactions in the bond market as a whole, and the turnover of repos accounted for 87.1 percent of the total transactions in the bond market. In 2006, the shares of these two turnover categories in the interbank bond market rose to 98.4 percent and 94.4 percent, respectively. Against this background, China's bond market can be seen as following the same momentum as the international bond market; the OTC market has already become the mainstay of China's bond market.

Fifth, a long-term mechanism for innovation in financial markets needs to be established. Innovation should be gradual and at a reasonable pace. Currently, China's legal system is still imperfect, and market participants have unsound corporate governance and relatively low capabilities in terms of risk prevention. Therefore, financial innovation and market construction should also follow a more gradual approach. The establishment of each component of the financial market system and introduction of new products should move ahead according to the sequence of "easy things first." China's experience over the past two decades has demonstrated that such a gradual course of development is helpful for ensuring stability in financial markets.

Institutional innovation, organizational innovation, product innovation, and technological innovation are all components of financial innovation. During the initial stage of market development, priority should be given to the construction of market institutions and organization. When a market has developed to a certain extent, focus can then shift to product and technological innovation. It is also important to emphasize the long-term innovation mechanism. So far, China has adopted a government-driven innovation paradigm. Such a paradigm has huge advantages in promoting the rapid development of a lagging market, because the government can mobilize all kinds of resources in the society to improve the legal system and infrastructure within a relatively short period of time.

At the same time, thanks to the government's strong monitoring and regulations, the development of financial markets has been smoother than expected. However, innovation ultimately needs to satisfy the market's needs. It is unavoidable that the government-driven innovation approach might fall short of meeting the needs of market players. Therefore, when the market has developed to a certain extent, the government should gradually step away from the forefront of market construction and let the market play a key role in innovation. The government's major role should be transformed to oversight for market fairness and transparency. Emphasis on developing market institutions and fostering the indigenous innovation capability of market participants will enable market competition and innovation to be carried out on a well-established and regulated platform and to move toward a virtuous cycle.

Conclusions

It can be seen from the latest developments of China's bond market that innovation is the driving force of development. Innovation includes the introduction of new bond products and new trading modes and, more importantly, the establishment of new institutions. Over the past three years, the bond market has achieved rapid development driven by product innovation and institution building. However, in the bond market, the bond products are still primarily Treasury bonds, central bank bills, and financial bonds; the development of financing bills still lags behind. In a glimpse into the future, the development of China's bond market still faces new opportunities. Great efforts need to be made in developing the bond market and expanding the scale of corporate direct financing. The development of the bond market has never been given such a priority. It can be predicted that China's bond market will certainly achieve rapid development if great efforts are made to promote bond market innovation and improvement in market institutions.

Note

This article has been revised on the basis of the author's speech at the Asia-Pacific Finance and Development Center (AFDC) biennial forum "Innovation for Development," Shanghai, September 21–22, 2006, jointly sponsored by AFDC and the World Bank.

References

Bond Research Society, Central Bond Registry and Settlement Co. Ltd. 2005. *Theory and Practice on Innovation in the Bond Market*. Beijing: China Market Press.

Department of Financial Markets, People's Bank of China. 2006. "Report on the Development of China's Financial Markets 2005." [In Chinese.] 1st ed. Beijing: China Financial Press.

Journal of China Finance. 2006. "Review and Prospects of the Development of China's Bond Market 2005." [In Chinese.] Special issue, *Journal of China Finance*.

Mu Huaipeng. 2005. *Asset Securitization in China: From Theory to Practice*. [In Chinese.] 1st ed. China Financial Press.

———. 2006. "Product Innovation and Institutional Building on the Inter-Bank Bond Market." [In Chinese.] Special issue, *Journal of China Finance*, April 4.

5

Innovation and the Financial Sector: Role of the Asia-Pacific Economic Cooperation

Kenneth Waller

Innovation and technology have been strong forces for development and transformation in the finance sector (Merton 1995b). Creative ideas and the application of technology are helping shape dynamic finance markets and mechanisms. Innovation can take a number of forms in the finance sector. It can entail improvements in the application of technology to transform and create new processes and improve efficiency, improvements in the quality of existing products and services, the creation of new products, and the creation of new markets. Some key aspects of the finance sector that have been enabled by the introduction of innovation and technology are summarized in the first section of this chapter.

The advances in finance identified in the first section have enabled and encouraged investment in technology and innovation in other sectors across an economy. This influence can be attributed to strengthened capacities in risk management, financial modeling, product design, and proliferation of service forms. These key strengths promote technology and innovation, both at the firm level in real sectors and through influences on specific markets for technology and innovation. The section is followed by a brief description of the role the finance sector can play in promoting innovation and technology.

The chapter then addresses the broader policy environment that promotes financial innovation. Such transformation requires financial stability but also requires exposure to competitiveness and openness as the key drivers for greater sophistication and efficiency. There are a number of preconditions for a successful transformation. These include strong legal, supervisory, and regulatory regimes, as well as a sound macroeconomic framework. A number of specific financial prerequisites are identified as key strategic challenges being tackled by Asia-Pacific Economic Cooperation (APEC) economies.

Finally, the chapter looks at the role regional cooperation can play in harnessing and facilitating innovation in the finance sector. There are a number of regional cooperation channels through which APEC and its affiliated bodies have generated this momentum in the Asia-Pacific region, with the higher objective of promoting sound and competitive financial systems.

Innovation, Technology, and Financial Market Development

The institutions, infrastructure, operations, and flows of finance are deeply influenced and facilitated by technology. Technology is a crucial "financial enabler," both enriching physical and human assets and driving productivity growth. Furthermore, creative ideas and new ways of thinking about finance have continuously pushed systems to become deeper, more efficient, flexible, and responsive (Merton 1995a). Evidence for this is in the great strides that have been made in the past three decades in areas such as banking, financial intermediation, insurance, risk pooling and management, and the enormous growth in value exchange across borders. More recently, innovation and technology have transformed international finance, facilitating the real-time transfer of wealth within and across economies on a massive scale (Merton 1995b). Some key aspects of the finance sector that have been enabled by the introduction of innovation and technology are identified in table 5.1.

Box 5.1 provides a brief example of how technology has transformed the functions of the banking sector.

How Financial Sectors Support the Financing of Technological Innovation

Developments in the finance sector, such as those described in the previous section, can affect the level of technological innovation in other sectors across an economy. In fact, the role of finance in fostering innovation and technology is considered one of the key mechanisms through which the

Table 5.1 *Key Aspects of the Financial Sector Enabled by Technology and Innovation*

Aspect	Advantage
Depth	Technology and innovation broaden the range and quality of products and services, risk management tools, and wealth creation. Such developments are evident in many sectors such as banking, securities, insurance, pensions and superannuation, and foreign exchange markets.
Efficiency	Electronic payments and transfers allow instantaneous and real-time transactions, providing significant efficiency gains and reduced transaction costs. Efficiency gains also come from a greater ability to reallocate wealth to the most productive activities.
Liquidity	Done electronically through new products, in new markets, wealth can be transferred quickly, efficiently, and at increasingly lower cost.
Mobility	Electronic payment allows for much greater mobility of finance and activities within and across national borders.
Risk management	Since the financial crisis of 1997, there has been recognition of the need for advanced sophistication in tools for risk identification, analysis, modeling, measurement, pricing, management, and risk diffusion. This aspect of the sector has seen developments in the areas of hedging instruments; in risk transfer markets such as securitization, derivatives, and options trading; and more recently in equity capital funding and collateral debt obligations.
Pricing	Information technology provides deeper access to information, greater communication, and more responsive pricing.
Sophistication	Greater information availability reduces reliance on intermediary activities for basic transactions, thus allowing for a focus of intermediation on more sophisticated complex transactions.
Choice	Innovative new products and services provide greater choice, allowing for a growing depth and responsiveness in the finance sector, as products are customized and tailored to meet customer needs.
Management and service culture	Innovation is also transforming the way institutions attract, communicate, and respond to their customers in an increasingly competitive environment.
Participation and financial inclusion	Electronically enabled markets, accessible information, greater choice, increased competitiveness, and improved risk management result in greater accessibility, increased participation, and financial inclusion. This aspect of the finance sector has expanded the volume of finance traded.
Regulation and supervision	Electronic data and analytics based on models enable new ways of facilitating transparency and accountability, regulating, and monitoring both by commercial financial institutions and financial system regulators.
Specialization	New products and markets allow institutions to specialize, such as in swap/contracting options, mortgage brokering, securitization, and hedge funds.
Microfinance	Innovation is fundamental to meeting the challenge of improving access to finance for the poor.

Sources: De Rato 2006; IFC 2006; Merton 1995a, 1995b; Reddy 2006; Su Ning 2006.

Box 5.1 *Technology in the Banking Sector*

Great strides have been made in enhancing the quality of banking in the Asia-Pacific region. This is reflected both in the improvement in commercial banking services and in the quality of banking supervision. Similarly, there have been major improvements in the infrastructure necessary to support efficient banking systems. Technology is helping to address key challenges in areas of security to complement the increasing reliance on information-technology driven systems, and it is fundamental to the operations of real-time clearing systems. Regionwide, bank customers access funds through automated teller machines, thereby allowing access at the time of a customer's choosing. Mortgage and personal credit approvals are often available online. Many banks in the region provide online banking services, enabling value transfers to nonbanking institutions. Digitized data transfers facilitate outsourcing of banking services to lowest-cost centers, which are often offshore from a bank's head office, providing an example of regional and global financial integration.

---

finance sector contributes and drives growth and total factor productivity in an economy (Benfratello, Schiantarelli, and Alessandro 2006).

The finance sector can support technological innovation through a number of general channels. As markets become increasingly more competitive and open, firms are constantly driven to become more innovative and efficient. For such firms, having access to external finance and the ability to raise internal finance are fundamental to financing technological and innovative advancements in their product markets (Ayyagari, Demirgüç-Kunt, and Maksimovic 2006). Empirical analysis shows that suppliers of capital play a significant role in encouraging innovation in an economy through their ability to pick winners. In other words, by developing sophisticated ways of seeking high returns, investors and finance institutions select successful potential projects, entrepreneurs, or businesses to finance, or potential markets to enter into, and then provide the finance required (Benfratello, Schiantarelli, and Alessandro 2006; Gompers et al. 2006; King and Levine 1993).

More sophisticated finance also provides opportunity for more complex investment endeavors, such as in markets for technology and innovation. Table 5.2 highlights some of the developments in finance that lend themselves to technology and innovation financing.

Table 5.2 *Financing Technology and Innovation: Some Characteristics of Innovation and Financial System Responses*

Characteristic	Financial response
Can be high risk	Risk dispersion.
	Improved risk management and capital allocation.
Can be capital intensive	An ever-expanding base of available capital to finance innovation, resulting from the development of globalized finance markets.
	Enhanced participation in financial systems (investors, savers, consumers, and producers) as finance markets, through innovation, become more accessible to a broader range of investors and savers.
	Increased confidence to participate in the economy, resulting from greater access to information and legal infrastructure, insurance, greater transparency and accountability, better-understood risks and availability of risk management techniques, and free and competitive markets. These factors increase the capital resources available, provide greater incentive for entrepreneurship, and provide more stable and long-term capital flows.
	Increased liquidity of finance markets.
Can be long term	Development of larger funds that pool resources, resulting in larger-scale and longer-term investment opportunities (such as public-private partnerships) for infrastructure investment.

Source: Author's compilation using sources cited in this chapter.

Policies and Best Practices That Encourage Financial Innovation

The previous sections have noted how innovation is intrinsic to the development of sophisticated financial systems and how, in turn, advancements in finance foster technological and innovative change in other sectors. Given the profound importance of innovation in finance, it is important to understand the factors that nurture it. This section examines the policies and best practices that encourage financial innovation.

Several fundamentals underlie the creation of an environment conducive to innovation in the finance sector. These are the preconditions for financial system stability, depth, and soundness, and they determine the quality of institutions and markets, and the infrastructure necessary for efficient financing. Indeed, many of these conditions have been identified as key strategic challenges that are being tackled by APEC economies. The preconditions include, but are not limited to, the following:

- A sound macroeconomic environment
- A fair, impartial, and enforceable legal framework
- A sound supervisory regime that encourages prudent risk taking
- A culture of good governance
- Sound risk management capacity
- Market discipline through good governance, but also through open, competitive, efficient, and accessible finance markets.

This section considers these preconditions in turn.

Sound Macroeconomic and Policy Environment

A sound macroeconomic environment promoting sustainable growth, price stability, and exchange rate stability can reduce a financial system's exposure to risk. Fiscal discipline that involves sensible and well-communicated policies is important for investor confidence and stable capital flows, both domestic and foreign. Flexible exchange rate policies facilitate smooth macroeconomic adjustments and can reduce exposure to sharp fluctuations in values. Options provide the means to hedge against currency risks. Sound monetary policy supports low inflation and price stability. An essential ingredient to the investment function is investor confidence, such that returns reflect risk and are not devalued by price inflation. Returns can be safeguarded by market-based risk techniques, including options and hedging to mitigate interest rate and exchange rate fluctuations. Such a policy environment reduces volatility and uncertainty and allows financial institutions the space to take risks and create. This in turn creates an enabling environment for risk taking by firms in physical markets.

Legal Framework

Underlying all economic interaction is an established, understood, and enforced set of laws or social rules. David Hume identified three basic principles that are as relevant in today's sophisticated and globalized finance markets as they were in the 18th century. These are the right to property, the conditions by which rights may be transferred freely, and the honoring of agreements on the transfer of property. These rules provide confidence and security to participate in the exchange of goods and services; they provide security under law to enter into commercial contracts. Relevant laws, principles, and standards that underpin transactions deal with the following:

- Private property
- Contracts and contract enforcement
- Corporate and bankruptcy laws
- Accounting, auditing, and disclosure of banks, nonbank financial institutions, companies, and public policy
- The empowerment of financial system regulators.

Security and consistency are paramount, as are principles of fairness, expeditious procedures, compliance and enforcement, and impartiality. These are necessary to ensure confidence in the system and, ultimately, participation. APEC member economies have identified the harmonization of legal procedures and practices as a key objective in attracting investment. Innovation is encouraged by the right to property and the right to the protection of intellectual property, which is frequently at the source of innovation.

Good Governance and Sound Regulation and Supervision

Regulation and supervision are fundamental to promoting efficient financial institutions, minimizing failure and insolvency, providing for smooth entry and exit of firms, and mitigating disruptions. Regulation and supervision provide the external frameworks for legal compliance, good practices, accountability, and governance.

Internal governance in firms is also paramount, and governance structures are the cornerstones of commercial financial institutions. Investors and shareholders require not only that their interests are protected under the law but also that boards and management structures act in ways to fully protect shareholder and depositor interests and to promote self-discipline. Similarly, integrity and trust in the role of the supervisory and regulatory institutions are critical preconditions of a sound financial system.

To develop a culture of good governance in both public and corporate institutions, the system must embody full disclosure of and access to information, prudential risk management, integrity, transparency and accountability, and competition and market discipline. APEC's strong advocacy of good governance in the region's financial systems is described in box 5.2.

Box 5.2 *The Rationale behind APEC's Focus on Good Governance, Supervision, and Regulation*

The promotion of good governance practice within APEC economies has been a priority of APEC and its members for much of the past decade. There are a number of reasons for this. First, the Asian finance crisis of 1997 highlighted (1) the need for enhanced and robust financial standards for the running and supervision of institutions, and (2) the role good governance should play in promoting financial stability. Second, there is greater interaction and interdependency of financial systems between economies, and good governance promotes the transparency, accountability, and trustworthiness required to generate the confidence of participants in cross-border economic and financial transactions. Finally, to attract investment, a financial system has to inspire confidence and security through governance principles of transparency and accessibility of information, good management, soundness, and policy stability.

Effective supervision and regulation in the finance sector is an essential complement to sound corporate governance arrangements in financial institutions, but not a substitute. Under the new Basel II arrangements, banking supervision is primarily focused on the supervisor assessing the quality of governance arrangements in a bank and the bank's effectiveness in managing risk and allocating capital according to sensible risk practices.

Given the globalized and integrated nature of finance markets, cross-border and cross-cultural interactions often occur. As such, moves to adopt (1) agreed-upon best international standards and practices in both regulatory agencies and in financial institutions themselves and (2) agreed-upon international accounting standards are important, both in ensuring stability and the success of increasingly integrated financial systems and in developing common approaches to governance arrangements that resonate across the APEC region.

Source: Author's compilation.

Risk Management

In the region's recent experiences with financial crises, bank failures caused by poor internal risk management in the institutions and poor governance have been common factors (De Rato 2006). Prudent risk management is vital to survival in a competitive and open environment, to the efficient allocation of financial resources, and to the achievement of financial stability. Prudential risk management requires institutional capacity to identify and understand risk, appropriately price risk capital, and manage and mitigate risks, all with the objectives of efficiently allocating capital and promoting investor confidence. Competition and good governance arrangements complement these objectives.

Innovation in the finance sector has helped develop sound risk management techniques and analysis, the capacity to better understand markets and trends, and the ability to provide for the application of technology and to produce efficiency gains. An increasing number of sophisticated risk management modeling tools are available to banks to help differentiate risks in particular asset classes and in different geographies. Hedge funds have expanded risk transfer markets by offering liquidity. Wide risk dispersion can increase the resilience of the system and the flexibility to adjust and absorb shocks (De Rato 2006). Notwithstanding these positive developments, risk diffusion can involve attendant problems, as evidenced by recent subprime mortgage defaults and the impact these conditions are currently having in international finance markets as a consequence of risk associated with collateral debt obligations.

Open, Innovative, and Competitive Financial Markets

A key determinant for market discipline is open, competitive, efficient, and accessible finance markets. Market discipline strengthens incentives for prudent risk management and is fundamental to the adoption of principles of good governance, efficiencies, innovation, and regulatory reform. A competitive system centers on choice. Empirical evidence shows that competition is a driver of innovation in finance sectors. A competitive system encourages knowledge transfer and synergies, and drives finance providers to find innovative methods, products, and services to remain competitive (Ayyagari, Demirgüç-Kunt, and Maksimovic 2006; Benfratello, Schiantarelli, and Alessandro 2006; Su Ning 2006). As noted in the preceding paragraph, risk diffusion can affect perceptions of financial market stability. It is outside the scope of this chapter to discuss issues relating to subprime mortgage markets. What is important to note is that markets are repricing risk, and adjustments, through risk premiums, are being affected. It is reasonable to conclude that good-quality borrowers for innovative products will continue to attract financial resources.

Importance of Regional Cooperation and the Role of APEC

Regional cooperation and collaboration can be crucial in harnessing and facilitating innovation in the finance sector. There are a number of regional cooperation channels through which APEC and its affiliated forums are generating change in the Asia-Pacific region, with the higher objective of promoting sound, stable, and competitive financial systems. This occurs predominately through capacity-building and information-sharing initiatives. Box 5.3 describes the membership, goals, and major objectives of APEC.

Box 5.3 *Background to APEC: Members, Pillars, and Objectives*

APEC comprises 21 diverse economies. There are high-, medium-, and low-income economies. There are capital-rich and highly innovative societies, and there are also less-developed ones. Each member has unique economic structures and institutions. The differences between members promotes opportunities for cooperation and collaboration, with great potential benefits to all involved. Membership in APEC is voluntary and nonbinding, again encouraging cooperation and collaboration to achieve regional objectives. The figure below depicts the APEC pillars and objectives.

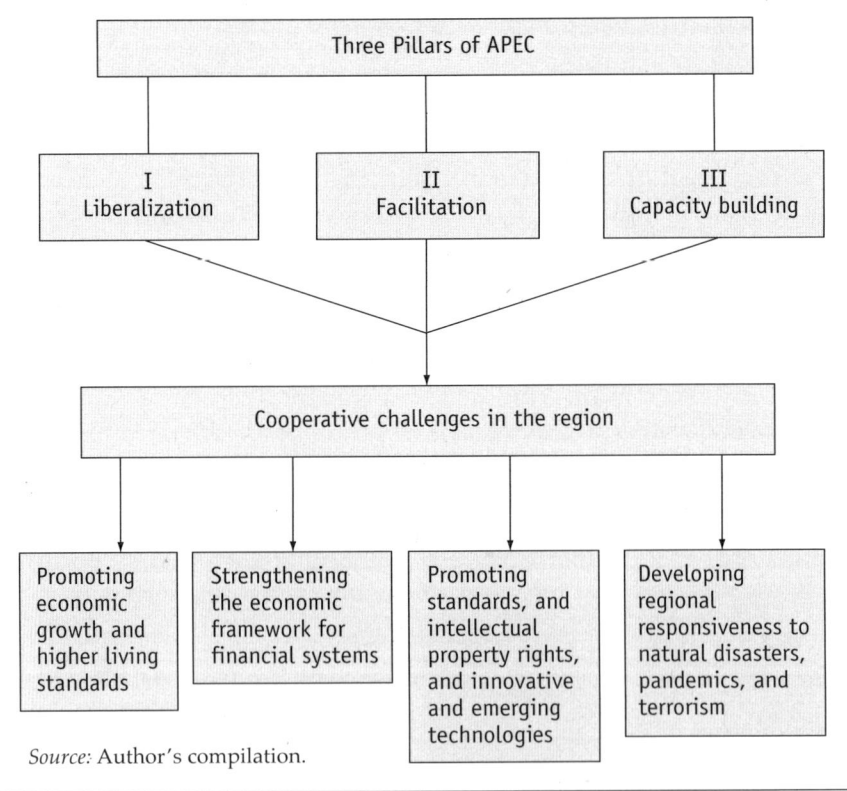

Source: Author's compilation.

Key Bodies and Their Activities in Promoting Financial Sector Innovation

Promotion of finance sector innovation and the development, strengthening, and deepening of finance sectors are complementary objectives. Competitive and sound financial markets promote innovation in financial services and encourage the creation of new ways of doing things—innovation and technology are fundamental components of deepening and broadening finance sectors. As such, promoting innovation is deeply interlinked with two of the main activities of APEC. The first involves advising member economies on how to enhance investment environments. The second is to promote deepening and strengthening of the region's finance systems. To these ends, APEC primarily contributes to finance sector innovation by promoting the prerequisite conditions outlined earlier in this chapter. In particular, APEC promotes competitive markets in which innovation and reward for risk are most likely to be fully compensated. APEC also works toward enhanced legal systems, including cross-border arrangements to protect creditor rights, and to strengthen and deepen capital markets and banking systems. APEC continues its work in promoting the development of bond markets, and specifically corporate bond markets.

The APEC finance ministers' agenda focuses on measures to strengthen the macroeconomic framework and to deepen and make more stable the region's financial markets. The Economic Committee is developing a structural reform agenda that includes enhancing competition and legal frameworks and governance arrangements. There is also an Investment Experts Group, which seeks to enhance local, domestic, and cross-border investments in the region. The APEC Business Advisory Council (ABAC) is the advisory body to the economic leaders of APEC on business matters in the region. Significant collaboration occurs among these various groups, central objectives of which include the promotion and development of innovative financial systems. APEC has also developed initiatives to address the challenges relating to securing information and communications technology. Furthermore, APEC is encouraging the development of regulatory and supervisory frameworks for financial markets and finance institutions in the emerging markets in the region in areas such as pensions, funds management, asset management, and bond markets. Finally APEC and ABAC support sector innovation by encouraging investment liberalization through collective and unilateral policies.

Capacity Building as the Primary Vehicle

Capacity building is one of the three pillars of APEC and is the primary vehicle through which APEC supports member economies in financial system development. Capacity building is a critical pillar to improve human capital and institutions, and it takes on a variety of activities and approaches to skills and knowledge transfer and development. APEC involves relevant institutions in the process of knowledge and experience sharing among member economies. These participants include multilateral institutions such as the International Monetary Fund, World Bank, Organisation for Economic Co-operation and Development, and Asian Development Bank; banking and professional groups; academics; and relevant regulatory agencies from the region's economies. APEC works with these groups to enhance awareness and to share experiences on the development of policy frameworks and regulatory and supervisory institutions.

ABAC and the Pacific Economic Cooperation Council (PECC) established the Advisory Group on APEC Financial System Capacity Building in 2003. The following are objectives of the group:

- To promote public-private partnerships involving business, national policy and regulatory agencies, and relevant international financial institutions and organizations
- To promote discussion and training aimed at enhancing regulatory supervisory capacities in various financial sectors
- To improve governance in both public and private institutions.

Strong public-private partnerships in the region have been established with groups such as the South East Asian Central Banks (SEACEN) Research and Training Centre and the Asian Bankers Association (ABA) to promote Basel II in the region's banking systems. Some of the key findings of a public-private dialogue held in 2006 (the second in the series) on the implementation of Basel II and developments in banking in the Asia-Pacific Region are described in box 5.4. Action to implement the findings would contribute to strengthened banking systems, to environments conducive to financial innovation in the region's financial systems, and, through that process, to more rational risk taking and innovation generally in the region.

ABAC works in partnership with regional and international groups, regulatory agencies, and finance market specialists in developing regional

Box 5.4 *ABAC Support for the Implementation of Basel II and Developments in Banking and Supervision in the Asia-Pacific Region, 2006*

An intensive public–private sector dialogue took place in Kuala Lumpur in 2006 between regulators, bankers, and international organizations organized by the South East Asian Central Banks Research and Training Centre, the Asian Bankers Association, the Pacific Economic Cooperation Council, and ABAC. Some key issues identified for further consideration are set out below:

- Major regional and multilateral agencies and economies should consider ways to intensify monitoring and surveillance and promote better disclosure of data on major imbalances that might affect regional and global financial stability.
- International banks should undertake improved reviews of their branches' and subsidiaries' cross-border operations, to ensure that the appropriate infrastructure is in place to implement Basel II and that local boards are well informed and have the expertise to meet governance obligations.
- Banks and supervisors should carefully review the appropriateness of risk models they may be employing to ensure that they are compatible with the broad objectives of enhanced risk management and governance, namely, to equip them to deal effectively with current and future stresses. Greater efforts are needed to deepen communication between home and host supervisors through supervisory colleges. Particular attention would be given to the validation of capital required by a host supervisor and to ways to make definitions more transparent, as well as to clarify both the process of approval and the supervisory reviews relating to capital. Bilateral and multilateral co-operation between supervisory agencies should be emphasized.

There remains an urgent need to undertake regional capacity-building activities between economies and their regulatory agencies, financial business groups, and regional and multilateral agencies; to develop best practices that assist in the cross-border implementation of Basel II; to develop financial consumer protection frameworks; and to promote good governance.

Source: Advisory Group on APEC Financial System Capacity Building 2007.

bond market initiatives and promoting a regional credit rating culture. Box 5.5 describes the most recent developments arising from a May 2007 public-private capacity-building dialogue to promote bond markets for corporate issuance in the APEC region.

Box 5.5 *APEC Support to Corporate Bond Market Development, 2007*

In May 2007, ABAC and the APEC finance officials held the first of a series of public–private dialogues on corporate bond market development in the region, focusing on Indonesia, the Philippines, and Vietnam. The dialogue between representatives from APEC economies, private sector market players, and experts from international public and private organizations had the following major objectives:

- Promoting public–private sector collaboration in the development of bond markets.
- Identifying aspects in the policy and regulatory areas of Indonesia, the Philippines, and Vietnam that could be addressed by authorities to enhance the environment for bond market development and, in particular, corporate bond issuance.
- Identifying capacity-building initiatives to build an environment that is conducive to bond market development. These might include public–private partnerships and regional cooperation initiatives.

The focus was on market development, market infrastructure, regulatory and supervisory environments, operational aspects of bond markets, the role of credit rating agencies, and the infrastructure to support markets.

A number of key capacity-building needs were identified at the forum that will shape APEC and ABAC agendas in the coming years, including the following:

- Expanding the institutional investor base through the development of pensions, insurance, and mutual funds
- Developing a strong credit rating industry in the region involving many actors such as the Association of Credit Rating Agencies in Asia, Finch Ltd., and experiences from other economies
- Promoting effective domestic and regionwide insolvency and creditor rights systems
- Promoting regionwide convergence toward robust global accounting standards.

Source: Waller and Parrenas 2007.

Finally, APEC works with the APEC study centers in the region and with training centers such as the Asia-Pacific Finance and Development Center (in Shanghai) and the Melbourne APEC Finance Centre, to undertake research, promote dialogue, and promote training programs for regional regulators in financial regulatory matters.

Conclusion

There have been recent and profound advances in the theoretical and empirical understanding of how advancements in financial systems facilitate growth and development in an economy. The application of technology and innovation to finance is driving many of these developments in the financial systems. In the past 30 years, technology and innovation applied to the field of finance have been a crucial "financial enabler," creating more efficient, flexible, responsive, sound, and dynamic finance markets and mechanisms. It is particularly relevant to look at the evolution of innovative finance techniques to meet the unique nature of investment in technology and innovation within firms and within innovation and technology markets.

A number of underlying preconditions contribute to creating an environment conducive to the adoption and promotion of innovation in the finance sector. Ultimately, they are important conditions for growth and development of an economy. In particular, the preconditions include a sound macroeconomic environment; a fair, impartial, and enforceable legal framework; strong supervisory and regulatory arrangements, good governance, sound risk management practices, and capacity; and competitive open markets. Many of these areas have been identified as key strategic challenges for APEC.

Regional cooperation can play a significant role in harnessing and facilitating innovation in the finance sector. There are a number of regional cooperation channels through which APEC and its affiliated forums are helping to enhance the environment for innovation. The critical challenge, and one that the region is moving toward in a most positive way, is that of promoting sound, stable, and competitive financial systems. APEC's regional capacity-building and information-sharing initiatives are highly supportive and are helping to create the environment that encourages private sector innovation.

References

Advisory Group on APEC Financial System Capacity Building. 2007. "The Implementation of Basel II and Developments in Banking and Supervision in the Asia-Pacific Region: Report of a Public-Private Sector Dialogue." Report to the Advisory Group meeting, AGFSCB 27-009, Seattle, WA, February 28, 2007. http://www.abaconline.org/v4/download.php?ContentID=3738.

Ayyagari, M., A. Demirgüç-Kunt, and V. Maksimovic 2006. "Firm Innovation in Emerging Markets: Role of Governance and Finance." Policy Research Working Paper 4157, World Bank, Washington, DC.

Benfratello, L., F. Schiantarelli, and S. Alessandro. 2006. "Banks and Innovation: Microeconometric Evidence on Italian Firms." IZA Discussion Paper No. 2032, Institute for the Study of Labor (IZA), Bonn, Germany. Available from SSRN, http://ssrn.com/abstract=848950.

De Rato, R. 2006. *The Growing Integration of the Financial Sector and the Broader Economy: Challenges for Policy Makers.* Madrid: Colegio de Economistas.

Gompers, P., A. Kovner, J. Lerner, and D. Scharfstein. 2006. "Skill vs. Luck in Entrepreneurship and Venture Capital: Evidence from Serial Entrepreneurs." NBER Working Paper 12592, National Bureau of Economic Research, Cambridge, MA.

IFC (International Finance Corporation). 2006. "Pushing the Boundaries of Microcredit and Peace to Innovations 'Beyond Credit.'" Retrieved March 4, 2007, from http://ifcblog.ifc.org/emergingmarketsifc/2006/10/microcredit_and.html.

King, R. G., and R. Levine, 1993. "Finance and Growth: Schumpeter Might Be Right." *Quarterly Journal of Economics* 108 (3): 717–37.

Levine, R., N. Loayza, and T. Beck. 2000. "Financial Intermediation and Growth: Causality and Causes." *Journal of Monetary Economics* 46: 31–77.

Merton, R. 1995a. "A Functional Perspective of Financial Intermediation." *Financial Management* 24 (2, Summer): 23–41.

———. 1995b. "Financial Innovation and the Management and Regulation of Financial Institutions." NBER Working Paper 5096, National Bureau of Economic Research, Cambridge, MA.

Reddy, Y. V. 2006. "Use of Technology in the Financial Sector: Significance of Concerted Efforts." Address at the Banking Technology Awards Function, Institute for Development and Research in Banking Technology, Hyderabad, India, September 2, 2006.

Su Ning. 2006. "Press Ahead with Reform and Opening-Up and Promote the Rapid and Healthy Development of the Financial Sector." Address at the Financial Summit of the Ninth China Beijing Hi-tech Expo, Beijing, May 23, 2006. http://www.bis.org/review/r060720e.pdf.

Waller, K., and J. C. Parrenas. 2007. *The First APEC Public-Private Sector Forum on Bond Market Development—Developments in Indonesia, the Philippines, and Vietnam and the Challenges and Prospects for Capacity-Building and Public-Private Partnership.* Conference Report of the Joint Forum organized by the Advisory Group on APEC Financial System Capacity Building and the APEC Business Advisory Council, May 7–8, 2007, Melbourne, Australia. http://www.oecd.org/dataoecd/29/5/40078541.pdf.

PART III

Innovation and Government Policies in China

6

Overview of China's Enterprise Innovation: Progress, Challenges, and Policy Recommendations

Chen Jia

Since the beginning of the 21st century, globalization and the development of advanced technology have led to dramatic changes in the world's industrial structure and division of labor. Knowledge-based economies—composed of industries such as information, software, and knowledge-based services—have become the primary drivers of world economic growth. Furthermore, economic growth has shifted away from reliance on natural resources and capital to the innovation of knowledge and technology. In light of these changes, developed countries have focused on the development of advanced manufacturing sectors within their countries, relocating less-advanced manufacturing and processing operations to developing countries through foreign direct investment (FDI) outflows and multinational corporations' mergers and acquisitions. As a consequence, labor divisions, particularly between skilled and unskilled workers, have also changed across countries.

Adjustments in industrial structure throughout the world present both challenges and opportunities for China's economy. In light of these changes, China's economic objective should be to harness technological innovation as the primary force behind its long-term industrial development.

To do so, it must upgrade its industrial structures and promote greater scientific and technological progress.

Since the start of China's open-door policy in the late 1970s, the economy has grown rapidly and has established a wide range of industries, many of which are ranked among the best in the world. However, China's economic growth is characterized by high levels of resource consumption and waste, environmental degradation, and low-value-added production. Encouraging indigenous innovation among China's industries is essential, not just to upgrade the technology of its industries to address these problems, but also to improve its international competitiveness.

The 5th Plenum of the 16th Central Committee of the Communist Party of China (CPC) recommended that China "increase the capacity for indigenous innovation to promote scientific and technological development, and transform its industrial structures."

Current Levels of Technology in Chinese Enterprises

Chinese enterprises have made significant progress in technological development since 1978. This progress has enhanced the capacity and level of technological innovation in enterprises, as described in the following sections.

Growth in the Level of Technology and Scale of Industry in China

The expansion of China's industries has led to an increase in its production capacity, which, for some sectors and products, ranks among the highest in the world. For example, China is now the largest producer of cement, chemical fertilizers, various steel products, chemical fibers, television sets, and motorcycles.

In addition to the great achievements in low- and medium-tech industries, such as clothing, textiles, and household electric appliances, China also has made great progress in high-tech sectors. For example, in the petroleum and chemicals sector, China's industry holds one-fourth of the world market and is currently third in the world in terms of output, with more than 10 kinds of products ranked among the top in the world in terms of market share. The shipbuilding industry also experienced tremendous growth, with output rising from 340,000 tons in the 1980s to as much as 8.8 million tons in 2004. The industry's share of the world market also rose from 1 percent to 14.6 percent during this time.

Table 6.1 *Share of Technology-Based Industries in the Manufacturing Sector, 1998 and 2003*
(percent)

Year	Low-tech industries	Low- to mid-tech industries	Mid- to high-tech industries	High-tech industries
1998	33.6	26.7	28.3	11.5
2003	29.3	26.9	29.3	14.5

Sources: Industrial Economy Research Institute of the Chinese Academy of Social Sciences; Lv Zhen et al. 2005.

China's industrial structure has also improved and shifted toward a more technology-intensive and more advanced structure. The share of high-tech industries in China reached 14.5 percent in 2003, up from 11.5 percent in 1998. In contrast, the value added for low-tech industries, as a percentage of total value added for the manufacturing sector, decreased from 33.6 percent to 29.3 percent (see table 6.1).

Changes in China's export composition reveal the upgrading of the country's industrial structure. Currently, electronics, mechanical, and high-tech products are China's main exports (see table 6.2).

Increases in Technological Capacities of Large and Medium Enterprises

China's capacity for technological research and development is improving. At present, 6,775 large and medium enterprises—23.7 percent of all China's large and medium enterprises—have established 9,352 research and development (R&D) institutions. The number of R&D personnel now

Table 6.2 *Changes in the Export Composition of China's Products, 1990–2005*
(percentage of total exports)

Exports	1990	1995	2000	2001	2002	2003	2004	2005
Primary products	25.6	14.4	10.2	9.9	8.8	7.9	6.8	6.4
Industrial manufactures	74.4	85.6	89.8	90.1	91.2	92.1	93.2	93.8
Electromechanical products	17.9	29.5	39.5	39.6	42.7	39.3	41.8	42.3

Source: National Bureau of Statistics 2006.

totals 1.7 million, of whom 1 million are scientists or engineers. In 2005, total R&D expenditure by large and medium enterprises reached 254.3 billion yuan (Y), a 209 percent increase from 2000. These figures demonstrate that large and medium enterprises are investing more in R&D in order to improve their technological capacities.

The number of patent applications filed by enterprises is increasing rapidly. In 2005, Chinese enterprises filed a total of 127,397 patents, which is 2.8 times the number in 2000. Of these, the number of patent applications for new inventions increased 4.8 times the number in 2000, to a total of 40,196 (see figure 6.1). The number of patent applications in China is on the rise. In 1995, China's high-tech industry filed a total of 612 patent applications. This number soared to 11,026 in 2004, an 18-fold increase in just 10 years. Patents for electronics, communications, and equipment manufacturing had the largest share of total patent applications in 2004, accounting for over 50 percent (6,986) of all patent applications among

Figure 6.1 *Patent Applications Filed by Chinese Enterprises, 2000 and 2005*

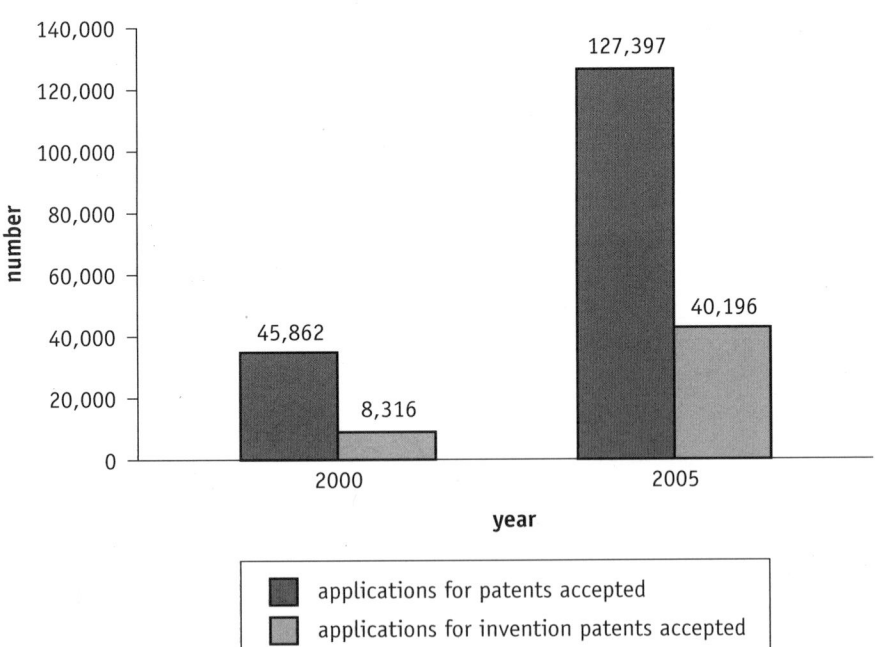

Source: National Bureau of Statistics 2006.

new high-tech industries. Pharmaceuticals were second, with 15.4 percent (1,696) of total patent applications.

A greater number of enterprises now have strong capacities for indigenous innovation. For example, Harbin Power Station Equipment, Dongfang Electric, and Shanghai Electric currently account for over 80 percent of China's production in power-generating equipment. In the steel and iron industry, Baosteel filed 349 patent applications in 2004—38.7 percent of the total number of applications by that industry—of which 109 were for new inventions. Patent applications were also outstanding for Haier Group: 5,469 of its patent applications were approved, of which 618 were for inventions. It also has 18 self-innovated core technologies. Another example of innovative capacity is Sinopec, in the petroleum and chemicals sector. Sinopec developed approximately 5,200 new scientific and technological findings and applied for 9,297 patents, both domestic and international. Seventy percent of the applications were for new inventions, and Sinopec acquired 5,199 patent licenses by 2007.

Growth of Small and Medium Technology Firms

Small and medium enterprises (SMEs) play a vital role in China's economic and social development as well as the enhancement of the country's innovative capacity. Currently, SMEs account for over half of China's GDP (measured in final products and services), 40 percent of its profits and taxes, 60 percent of its exports, and 75 percent of urban employment. Since 1995, the share of SMEs in high-tech industries has steadily increased, and SMEs are now one of the most dynamic segments of China's high-tech industries. Data from China's Ministry of Science and Technology indicate that over 90 percent of new high-tech enterprises are SMEs. SMEs also account for 65 percent of all invention patents and 80 percent of new products.

Private technology-based firms are one of the major drivers behind innovation and industrial development in China. These firms have developed rapidly, and their capacity for indigenous innovation has improved as a result. By the end of 2005, the number of private technology-based firms reached a total of 143,000—an increase of 1.9 percent since the previous year—and total assets also increased by 19.1 percent, to Y 6,312 billion. In terms of employment, these firms now employ approximately 1.85 million employees with college degrees or higher, accounting for 15.3 percent of the total workforce of private technology-based firms.

R&D expenditure of private technology-based firms has also increased, as has income derived from technology-based products and services. In 2005, R&D expenditure rose by 24.6 percent from the previous year, reaching Y 123 billion. Technology-related income also increased by 16.6 percent from 2004 to 2005, up to Y 275.7 billion (Ministry of Science and Technology 2006b).

Regional Growth of Industrial Clusters

Experiences of China and other countries highlight the role of industrial clusters in enhancing the innovative capacities of firms. Clusters can combine industrial development with the regional economy, taking advantage of industrial synergies and collaboration to increase innovation and development. Other advantages of industrial clusters include increasing labor specialization and improving the flow and allocation of technology, talents, and capital. The emergence of industrial clusters has had a significant impact on China's economic development and continues to influence the country's growth patterns.

Currently, China has several particularly strong industrial clusters in different parts of the country. In Zhongguancun, Beijing, new high-tech enterprises formed a cluster that is now one of the forerunners of China's high-tech industrial development and an example for other industrial clusters in the country. Clusters in the Yangtze River Delta and the Pearl River Delta are also very active in technological innovation, a result of proactive policies and efforts to promote indigenous innovation among their enterprises. For example, more than 300 out of 1,500 townships in the Guangdong Province have developed specialized local economies. The eastern coast of the Pearl River is home to one of the largest clusters of electronic information industries in the country, while the western coast is dominated by a large industrial cluster for household appliances.

Development of New High-Tech Industries

The size of new high-tech industries is steadily growing. In 2005, the total output of China's high-tech industries reached Y 3.44 trillion, an increase of 23.8 percent from the previous year. High-tech industries accounted for 15.5 percent of the total output. Since the start of the 10th Five-Year Plan, the output of high-tech industries has maintained a high rate of growth, av-

Figure 6.2 *Total Output Values of China's High-Tech Industries, 1995–2004*

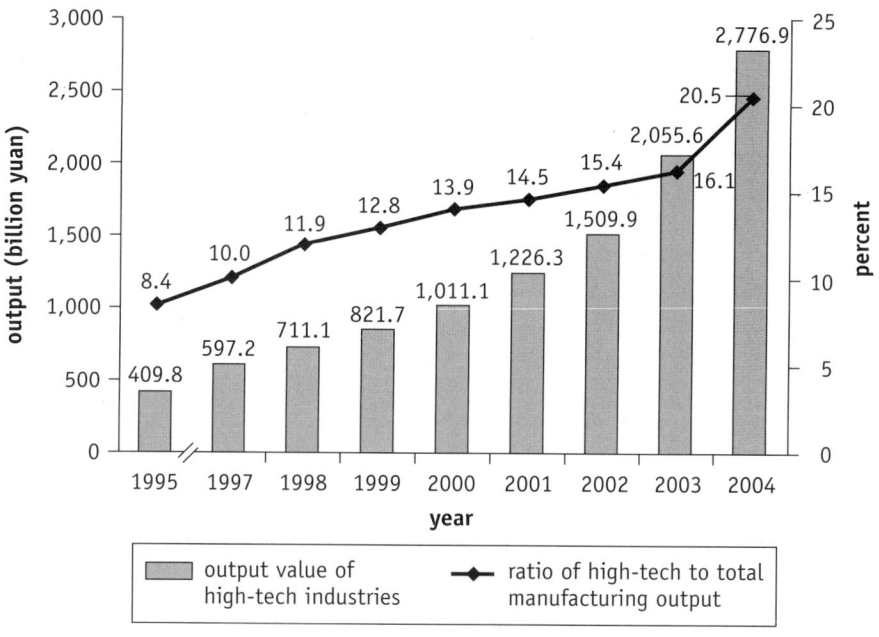

Source: Ministry of Science and Technology 2005.
Note: For technical reasons, the data for 1996 are not comparable.

eraging as much as 27 percent annually. This growth rate is 6.5 percentage points higher than the average growth rate of high-tech industries during the 9th Five-Year Plan and 3.3 percentage points higher than the growth of the manufacturing sector as a whole (see figure 6.2).

Problems Facing Indigenous Innovation among Chinese Firms

Although the levels of technology and innovative capacity among Chinese firms have increased greatly, Chinese industries as a whole still suffer from low innovative capacities when compared with advanced countries. Furthermore, data show that while small numbers of firms and industries have become internationally competitive, this is not the case for the whole economy. Improving the overall technological level and innovative capacity of China's industries is a problem that still must be addressed.

Weak Technological Development

The majority of China's industries are still low on the world's industrial value chain and have relatively low technological levels. Most of China's industries remain technologically underdeveloped when compared with industrialized countries. For example, the machinery industry often serves as an indicator of a country's level of industrialization and national strengths. However, China tends to produce mostly lower-level processed and ordinary products and is still unable to produce many key types of machinery and equipment. The country still lags noticeably behind advanced countries in the production and quality of important manufacturing equipment such as gas turbines, nuclear power generators, hydraulic power equipment, high-speed trains, light industry textiles, and clothing manufacturing equipment.

In the software industry, the level of system and supporting software is used to gauge the technological level of a country. In China, system and supporting software account for only 25.8 percent of all software products, whereas 74.2 percent is application software. This suggests that China's software industry is comparatively weak.

In shipbuilding, China lacks key technologies and continues to trail other countries that possess high-tech ship design capabilities. Japan and the Republic of Korea continue to be far ahead of China in terms of the technological levels of their shipbuilding industries. In fact, China's shipbuilding industry is approximately 10 years behind these countries, and ship-supporting industries in China fare even worse—about 20 years behind Japan and Korea.

Industries still suffer from low-value-added and poor economic returns. Data from China's State Statistical Bureau indicate that the present value-added ratio for the country's manufacturing sector is 26.5 percent. This is much lower than the manufacturing sectors of the United States (49.5 percent), Japan (48.5 percent), and Germany (38.2 percent). In the manufacturing of communications equipment, computers, and peripherals, China's value added ratio is only 22 percent—35 percentage points lower than advanced countries such as the United States.

China's Dependence on External Sources of Technology

For much of its industrial development, China used a strategy of opening markets to exchange technologies with other countries. However, in response to increased globalization, many countries have decided to increase the protection of their technology exports. Obtaining core technologies through the world market is now very difficult, and China oftentimes is unable to find or acquire the technologies it needs. China is also still subject to constraints posed by other countries that possess core technologies China requires for industrial development.

China lacks indigenous intellectual property rights (IPR) that would have serious negative effects on China's industrial development. In 2004, 63 percent of China's new patents for inventions were held by foreign individuals or firms, the vast majority of which were concentrated in high-tech industries. Data indicate that the United States and Japan hold approximately 90 percent of all high-tech patents. Consequently, the lack of IPR held by China's industries means that the country must rely on others for core technologies. This reliance creates a bottleneck for the further development of China's industries and their entry into the international market.

China depends on large quantities of imports for much of its high-end technical equipment. Although China is a major exporter for many industrial products, it still performs poorly in the production of many forms of equipment, especially technical equipment. The country's manufacturing sector is quite large, yet it lacks sophisticated processing techniques and key technical equipment that characterize strong competitiveness in the sector. China urgently requires more high-level technical equipment to produce high-tech products and increase the level of value-added in its products. Consequently, the country is dependent on imports to meet these needs, as China lacks the capacity to develop and produce technical equipment on its own.

Currently, two-thirds of equipment purchased in China is imported, and the annual trade deficit in equipment manufacturing products is in the tens of billions of U.S. dollars. Statistical data indicate that China depends on imports for virtually all of its needs for certain technical products, such as fiber-optic manufacturing equipment and control units for power generators. China also relies on imports for as much as 85 percent of integrated circuit chips, 80 percent of petroleum and chemicals sector equipment, and 70 percent of sedan car manufacturing equipment.

Weak Capabilities to Absorb and Adapt Imported Technologies

Reverse engineering and the absorption of imported technology are a shortcut for latecomer countries to accumulate and develop their own technological capacities and catch up with advanced countries. However, most Chinese firms do not take advantage of these opportunities. Absorption of imported technology is low, and seldom do firms conduct further innovation on imported technology. There are several causes behind China's low level of technological absorption.

First, China has low inputs into the digestion, absorption, and adaptation of imported technologies. In 2004, the ratio of financial inputs to technology imports as they relate to digestion and absorption of these technologies in large and medium Chinese industrial enterprises was only 1:0.15, while such a ratio in Japan and Korea had consistently been about 1:5.0–8.0. For example, in the sedan car industry, since the mid-1980s China has successively introduced foreign brands and equipment. Up to now, almost all the major sedan car brands around the world have been introduced into the Chinese market, and the ratios of domestic contents for medium- to low-range car models are mostly over 40 percent, with the figures for some car models as high as 70 percent or even over 90 percent. However, because of China's failure to undertake full digestion and absorption of imported technologies, core car-making technologies have been controlled by foreigners.

Second, more emphasis can be placed on hardware (manufacturing equipment) imports instead of software (technologies for industries) imports. A recent survey of Zhejiang Province shows that, among technology imports by large and medium enterprises, 93 percent of import contracts were equipment based, whereas only 7 percent were technology based. In contrast, equipment imports by Japan and Korea are only 23 percent of the total, and almost 80 percent of technology imports are software. Since Chinese enterprises import mainly equipment to expand production, they end up overlooking opportunities to obtain and absorb imported technologies.

Third, too much emphasis is placed on FDI inflows instead of technology. Many localities and enterprises are too focused on the amount of foreign capital inflows and neglect the importance of introducing new technologies. As a result, many foreign-invested projects are low in technological content. Some employ technologies with expired patents, some use outdated technologies, and others are even environmentally damaging. The lack of indigenous IPR and core technologies force Chinese enterprises to be on the lower end of the production and value chain of the global market.

Lagging Commercialization of R&D Results

Only 10–15 percent of China's R&D results are commercialized and achieve economies of scale, far lower than the rate of 60–80 percent in advanced countries. In recent years, the ratio of China's investments in R&D related to investments in commercialization of R&D results is between 1:1.0 and 1:1.5, far below the international average of 1:10. Insufficient inputs in commercialization have made it difficult for many R&D results to be put to practical use. For instance, the ratio of commercialization of petroleum and chemical industry R&D results is less than 30 percent.

Lack of Integration between Industry, Academics, and Applied Research Communities

Given that innovative capacities of China's firms are relatively low, it is vital for China to take advantage of the research capabilities of its scientific research institutes and institutions of higher learning to better integrate production and research. However, production is often still isolated from academia and research institutions. Consequently, the supply of and demand for technologies in China are not linked, so technology solutions offered by research institutes seldom meet the actual needs of enterprises. Additionally, enterprises continue to face problems of lack of access to sustainable and stable sources of technologies and talents.

Research institutions and universities obtain the vast majority of their funding from the government and very little from the business community. In 2004, 79.8 percent of R&D expenditure available to Chinese institutes of scientific research came from government funds, with merely 5.2 percent coming from enterprises. R&D expenditure of universities had a higher share of funds from enterprises, about 37.1 percent.

The majority of technology purchases by firms come from other enterprises, especially foreign-invested enterprises. In comparison, technology purchases from scientific research institutes and universities are very small, and their share is actually decreasing. The number of contracts between enterprises and research institutes as a percentage of the national total decreased from 27.5 percent in 1998 to 20.8 percent in 2004. Their share of total contractual value also declined from 34.9 percent to 14.3 percent. While the number of technology contracts involving universities has remained relatively stable, their share of total contractual value has also declined, from 11.9 percent to 8.7 percent from 1998 to 2004.

Reasons for Weak Indigenous Innovation among Chinese Enterprises

Previous sections have briefly analyzed the current situation and problems that exist in technological innovation in Chinese enterprises. The following sections relate to weak innovation in Chinese enterprises.

Lack of Integration between the Economy and Science and Technology

Although China's scientific and technological systems have changed dramatically in recent years, the country has not paid sufficient attention to incorporating science into its economic development. This is because scientific development and economic development have continued to be viewed as separate systems with their own paths. Consequently, few efforts have been made to promote synergy between economic development and scientific development or firm-led technological innovation.

Science and Technology (S&T) Policy China's scientific and technological sector looks primarily at patents and product prototypes as the primary outcomes of development. Commercialization and industrialization of R&D results play a far smaller role. Out of China's five priorities for science and technology, the only one that addresses the utilization of R&D results is the plan for building an environment for the industrialization of S&T results. In 2004, this plan received only Y 1.27 billion in allocations, less than 1.2 percent of the central government's total allocation for science and technology.

Economic Policy China's economic policy focuses primarily on cumulative effects of investment, consumption, and exports on GDP growth. Science and technology are not a major consideration and take a back seat. Furthermore, China's economic policy lacks clear goals and objectives for the utilization of scientific and technological results. Few economic policy components are designed exclusively to support the commercialization and industrialization of technology results derived from indigenous innovation.

The Innovation Process Before R&D results are commercialized, they undergo a stage of application development, whereby product prototypes are converted into mass production models. However, China's current policies place little emphasis and support on this stage of development. This is because there is no linkage between R&D results and actual industrial production. For example, China's laser industry has paid great attention to the development of laser application solutions. The Chinese Academy of Sciences alone has six institutes engaged in laser-related research; Tsinghua University and Peking University also have dedicated laboratories. The government has also invested large amounts of human, physical, and financial resources into these research institutes, and the R&D results have been successful and on par with advanced countries. Nevertheless, most of the results remain as certificates, theses, and prototypes because the research institutes are not responsible for introducing their results into the market. On the other hand, enterprises cannot conduct effective pilot tests of these new results because of limitations in funding, talent, and technology. This problem is exacerbated by the lack of government policies to aid in the conversion of R&D results to market. Consequently, China's demand for key laser equipment must be met instead by imports, even though the country possesses its own advanced laser technology.

The separation of China's economic and science policies is a symptom of larger systemic issues. To address these underlying problems, more reforms to promote scientific development are needed. Additionally, China must foster mutual support and integration between its scientific and economic systems.

Lack of Overall Deployment Strategy and Effective Policies for Indigenous Innovation among China's Important Industries

Government intervention is necessary to promote innovation and speed the development of China's strategic industries. However, government policies in this field are weak and uncoordinated, so they are largely ineffective. There are several causes for this problem.

Lack of an Overall Plan for Sector-Based Innovation and Technological Progress   In the 1990s, China's government implemented institutional reforms to reduce its control of state-owned enterprises through a policy of *zhuada fangxiao* (controlling the big while releasing the small). Consequently, many of the government's powers over the economy were diminished in order to promote market-based policies. However, these policies also reduce the scope of guidance and administrative functions that the government could use in the long-term development of the country's industrial sectors, such as industrial planning, operational coordination, and technology-development task forces.

For example, the development and production of large power-generation equipment often involves long R&D cycles and significant technological difficulties. However, China lacks an overall strategy to encourage indigenous innovation for the production of these kinds of equipment. This is partly due to a lack of incentives: the business community is satisfied with the level of technology currently being used. When new technology is developed overseas and newer generations of equipment are introduced, China is then forced to import the newer technology. Thus, instead of being forward thinking and developing their own technology, China's power-generation industries look only at short-term objectives and fail to sufficiently invest in innovation.

Nuclear power generation is also an area where China lacks a long-term development plan. China's nuclear power equipment manufacturing sector is at least one generation behind its foreign counterparts, largely due to the decade-long debate that started in the 1970s over the development of pressurized water reactors, which wasted precious time and hampered the industry's capacity to innovate and advance. Thus, when China began to develop nuclear power on a broad scale, the lack of technical equipment and capacity led it to rely on other countries to supply the needed equipment and expertise.

Another example is the aviation industry, particularly the manufacturing of large aircraft. Since the 1980s, China has debated whether to manufacture such aircraft, including what types of aircraft to produce and where they should be manufactured. Even now, there has been little progress or consensus in the debate. Consequently, the supply of aircraft equipment in China is almost entirely dominated by foreign suppliers.

Lack of Industry-Specific Technical Standards and Quality Controls   Industry-specific technical standards and quality controls for research have weakened, and many common technologies are neglected. Prior to the institutional reforms of the 1990s, China used to have over 300 industry-specific research institutes. The reforms led most of these institutes to be acquired by or converted into private enterprises. As a result, industry-specific research institutes that the government had invested decades of time and efforts into were abandoned or transformed to suit a different function. The research and technology sources that these institutes once provided are now no longer available, and this has hurt the level and availability of industrial technology for many key sectors.

Lack of Institutional Protection and Clear Plans for Advanced Technologies The government has weak regulatory control over mergers and acquisitions by foreign investors. Consequently, many enterprises that once played a key role in several industries have been taken over by multinational companies. Local brands and indigenous innovative capacity have declined, while dependence on foreign technology has increased. This problem poses a threat to China's industrial security and long-term interests.

Poor Protection of Intellectual Property Rights

International experiences indicate that protection of intellectual property rights plays an important role in promoting technological innovation in the industrial sectors. Although significant progress has been observed in IPR protection in China in the past decades, many efforts are still urgently needed to improve IPR protections. The relatively weak IPR protections in China now would be attributed to following aspects.

An Imperfect Legal System Since the establishment of its intellectual property rights regime in the 1980s, China has promulgated 15 IPR-related laws, regulations, and administrative decrees in order to im-prove the protection of IPR in the country. Compared with advanced countries, however, China's IPR regime is still weak. Intellectual property infringements are rampant because enforcement is lax and existing options to protect and enforce IPR are insufficient. For example, China has no antimonopoly rules for IPR and no restrictive measures to prevent IPR abuses. Additionally, many rules and regulations are unnecessarily restrictive. Patent notices, as stipulated by the Patent Law, take as long as 18 months, and examination periods for new invention patents can take as long as three years. This makes the patent-approval process take as long as five or six years to complete. These restrictions make it difficult for firms to innovate and they dampen incentives for technological progress.

Lax Enforcement China's IPR protection system involves both judicial and administrative activities that run in parallel. However, both aspects suffer from poor levels of enforcement. Legal provisions are often ignored, and lax law enforcement is commonplace. For administrative enforcement, local protectionism causes many administrative agencies to ignore legal provisions and fail to seriously crack down on infringement activities. In judiciary proceedings, court rulings often come very slowly, and litigations are time-consuming. Consequently, the effective level of IPR protection in China is actually quite low, despite having a relatively strong IPR protection regime on paper.

Lack of Scrutiny over IPR in Many Major Economic Activities   China lacks sufficient supervision, investigation, and prevention mechanisms to over-see many important economic activities, such as mergers and acquisitions by foreign investors or technology exports. In particular, attention is primarily paid to fixed assets such as land, machinery, or equipment. Non-physical assets such as brands, trademarks, and patents are consequently neglected.

Performance Indicators and Indigenous Innovation

Government authorities rely primarily on GDP as the main indicator of economic development and political achievement by public officials. At the same time, these authorities fail to pay sufficient attention to other important indicators, such as indigenous innovation among firms. Innovation as an indicator is also seldom included in state-owned enterprise (SOE) performance assessments by state asset management agencies. Currently, SOE performance is based on aggregate profits, ratio of net asset returns, and industry-specific criteria. The lack of innovation and technology development indicators reduces the incentives for firms to innovate and encourages short-term-oriented firm behavior. If these trends continue, firms will have fewer and fewer core technologies, as older technologies are replaced by foreign imports rather than developed domestically. Technological capacity will also decline, and China will lose its comparative advantage in several industries. To address this, the State-Owned Asset Supervision and Administration Commission has begun to include innovation indicators such as science and technology inputs in SOE performance criteria. Nevertheless, more effort is needed to include technology development and innovation indicators in overall performance measurement.

Weak Policies and Institutions to Support Indigenous Innovation

China has no policy to encourage the demand and consumption for new and innovative products. The majority of innovation policies are supply side: they look at science and technology inputs. Demand-driven policies to increase consumption and demand for innovation and new products are very rare. More incentives are needed to raise the demand for innovation.

The range of products permitted for government procurement is very limited. Under China's government procurement policies, most of the products entitled to procurement are consumer goods for daily use. In contrast, equipment and manufacturing products are generally not included under government procurement. This exclusion hurts the entry of innovative products into the market. For example, new innovative equipment often has no track record, so it is usually excluded from government procurement lists. This makes it difficult for these products to gain broader market access and consumer confidence. With the exception of a few major government-sponsored engineering projects, most new products face significant barriers to the market, and government procurement is unavailable for assistance.

No policies are designed to promote collaboration between industry and academia. China lacks the policies and institutions to address potential problems in industry-academic collaboration, such as issues of ownership, IPR protection, trade secrets, employee flows, and the distribution of benefits and liabilities resulting from collaboration. Consequently, if these problems arise, they are difficult to resolve; industries and research institutions are often reluctant to form partnerships because of the legal and financial risks involved. The lack of collaboration is a tremendous missed opportunity for China.

Policy Options to Enhance Indigenous Innovation among Chinese Enterprises

China's economy has steadily moved toward a market-based system since the late 1970s, and these changes have provided incentives for enterprises to innovate and become more dynamic than ever before. While China faces several obstacles that inhibit the innovative capacity of its firms, history and experience have shown that market mechanisms are the most appropriate way of assisting China's enterprises in promoting technological development and innovation. Thus, appropriate government interventions in order to correct market failures and protect the country's strategic interests are needed in order to best utilize the market.

Develop and Form National S&T Plans and Industrial Policies for Indigenous Innovation

A national science and technology plan incorporating several key projects provides a strong incentive for indigenous innovation and greater investment in R&D. The objectives for such a national S&T plan would include the following: (1) organize resources efficiently, (2) promote research collaborations between enterprises and academia for the development of advanced technologies, and (3) achieve significant technological breakthroughs in key sectors. There are several steps China should take to reform its industrial policy to promote innovation.

1. Efforts should be made to accelerate the technological development of traditional industries. The development of core technologies is one important aspect, but the government should also seek to accelerate the development and use of new advanced technology to restructure its traditional industries such as clothing and textiles. Prioritizing the development and use of information technology is one way to improve these industries.

2. China must take a long-term, forward-thinking approach toward science and technology by promoting the development of new, high-tech industries. Biotechnology and nanotechnology are two such fields for which early development is crucial. Early entry into these sectors would allow China to capture an early share of the market and gain a foothold in the development of these emerging industries, giving it a favorable position for international competition.

3. China should place more protection on its sensitive industries to prevent too much reliance on foreign firms. The *Industry Guidance Catalogue for Foreign Investment* must therefore be revised to place China's strategic industries in restricted categories. Provisions must also be made to limit the amount of control and ownership foreign investors can have of Chinese industries.

4. The government needs to leverage the public funds to induce more social investments into the R&D activities. This is particularly im-

portant for the development of modern service industries and information technologies. China should also seek to raise the share of its service industry relative to the entire economy.

5. Indigenous innovation and R&D among firms should encourage environmental protection and reduce the consumption of energy and resources. Restrictions should be placed on techniques and products that are inefficient and polluting. This will help encourage the development of more efficient technologies to ensure sustained long-term economic and social development.

6. The introduction of advanced technologies must also be accompanied by the absorption of such technologies so that firms can further improve their technological capacity and reduce their dependence on foreign know-how. Doing so would give Chinese sectors new technologies that are better designed for China's unique needs. Furthermore, greater technological absorption will also enhance the innovative capacity of China's firms and allow them to leapfrog in terms of technological development.

Strengthen IPR Protection

More efforts must be made to improve China's IPR protection regime and increase enforcement activities in accordance with the law. Additionally, other aspects of the IPR regime must also be looked at, such as regulations on mergers and acquisitions by foreign firms. The overall goal of IPR reforms is to improve the level of protection and enforcement while ensuring that the IPR system benefits China and creates incentives for Chinese firms to innovate. Greater guidance and coordination between various government authorities is essential to achieve these tasks.

Establish R&D Platforms for Common Technologies and Strengthen the Innovation Environment and Technology Standards

R&D for core technologies and strategic technologies involves national security and China's long-term economic and social plans. Priority areas for R&D are described in the "Outlined National Program for Medium- and Long-Term Development of Science and Technology (2006–2020)" (Ministry of Science and Technology 2006a). According to the national priorities defined in the plan, China should establish some key national laboratories and technology centers that meet international standards. These institutions are to address existing bottlenecks in technological development, perform covert basic research into core technologies, and assist in the application of R&D results in production and entry into the market. These laboratories and centers must be part of an integrated R&D platform to promote collaboration and boost innovative capacities.

Meanwhile, priority needs to be given to the development of the S&T infrastructure and to an environment conducive to indigenous innovation. The government should make proactive efforts to support enterprises by creating an environment that encourages firms to innovate rather than inhibits them. One step is the establishment of public R&D platforms to strengthen China's technological infrastructure. Other steps include the creation of additional science and technology parks and their supporting systems and infrastructure, stronger IPR protection, better product quality assurance mechanisms, and support for investment guarantees and other agencies offers of financial services for indigenous innovation.

It is also crucial to develop sound technical standards. These technical standards and regulations should guide both industry and academia to form partnerships to advance R&D activities.

Improve the Absorption and Adaptation of Introduced Technologies

Increasing the level of technology absorption is essential because it allows Chinese enterprises to redevelop existing technologies and reduce their foreign dependence. Technology introduction mechanisms should therefore be established to facilitate the introduction and absorption of new technologies. For example, large-scale national construction projects can serve as important catalysts for indigenous innovation by bringing in a variety of new, advanced technologies from abroad. These projects can then serve as a starting point for technological absorption and reinnovation, such as through reverse engineering, in order to achieve major scientific breakthroughs that would have been impossible without foreign technologies.

Promote the Commercialization and Industrialization of Technologies

Public spending on science and technology accounts for a large percentage of total S&T investments. Consequently, the results of public spending in R&D also belong to the state. However, the government lacks effective methods of technology transfer and dissemination, so many results and discoveries remain stuck within universities and research institutes and never make it to the market. In recent years, only 10–15 percent of all scientific and technological results in China have been commercialized—a level far lower than those in advanced nations. Financial investments, therefore, are unable to be turned into productive capacity and their social benefit is very low. China therefore needs to implement more effective policies to promote the dissemination and commercialization of newly developed technologies. China can take several steps toward this objective:

1. Establish a government technology transfer agency. This agency would be responsible for disseminating public and private information on newly developed technologies and aid in technology transfers between academia and the private sector. These agencies would also play an important role in monitoring and inspecting technological developments in order to ensure compliance with IPR rules and regulations. China would not be the first to establish these kinds of agencies. For example, the United States has several government technology application offices established in leading national laboratories. The U.S. Department of Defense, in conjunction with other departments, has also set up the Federal Laboratory Technology Transfer Alliance, and the Federal Technology Application Center is established under the Department of Commerce. These agencies are responsible for the provision of federal information related to technology transfer. The Japan Society for the Promotion of Science in Japan and the South Korean Technology Progress Company in Korea are also examples of publicly funded agencies responsible for technology transfers.

2. Clearly define scope and objectives for technology transfers in government-sponsored projects. Government-sponsored S&T projects need to include specific provisions on how the results of the project will be transferred or disseminated. Additionally, intellectual property rights also need to be clearly defined, and ownership of these rights should be specified before the project starts. Technology transfers, dissemination, and commercialization all need to be incorporated as criteria for performance assessments.

3. Introduce different management approaches for basic research and industrial technological research. For basic science and research, management should mainly focus on setting up a reward system

and sound measurement system for scientific papers. For R&D of industrial technologies, the focus should be on how to use the patent system to protect innovators' interests and the market mechanism to provide incentives for R&D.

4. Localize imported technologies. For emerging industries, the government should focus on importing and absorbing technologies, incubating enterprises, and developing new sectors. For traditional industries, innovation policies should focus on technological upgrading and introduction of process-improving technologies as well as the absorption and dissemination of technology.

Develop a Technology Innovation System with Enterprises as the Cornerstone and Close Links between Industry and Academia

To comprehensively enhance enterprises' indigenous innovation capabilities, the key is to strengthen the central position of enterprises in innovation. To that end, the system needs to take into account the laws and characteristics of technological innovation and to make greater efforts in reforming the S&T management system; optimizing the structure of S&T expenditure in public finance; changing the current model of putting most public R&D resources into institutions of higher learning and institutes of scientific research; giving priority to the creation of a technology innovation system with enterprises as the cornerstone with linkage among industry, universities, and research communities; and beefing up the government's role in guiding and mobilizing societal resources into S&T activities.

More specifically, the following actions are needed:

1. The formulation of national plans for science and technology should fully take into account the needs of industrial upgrading, technological progress, and productivity growth.

2. National plans should actively seek the advice and participation of the private sector. The government needs to establish means of communication and dialogue with enterprises to understand their needs and reorient its plans to benefit enterprises. Project appraisals and approvals should also solicit the feedback of enterprises. Alliances between enterprises and academia can also play a leading role in various special projects for industrialization. In sum, project design and implementation need greater collaboration between

government, enterprises, and academia so that plans are more effective and efficient.

3. More research projects conducted by research institutes and universities should focus on addressing technological barriers of production and business sectors. Government-financed projects should focus on these issues as well, but they should also give priority to applied technologies and difficult technology areas.

Support SME Innovation

SMEs are among the most dynamic and efficient enterprises in China's economy and are responsible for a large share of technology creation and scientific development. However, SMEs suffer from an unlevel playing field compared with large enterprises, which limits their access to resources and their capacity for innovation and R&D. SMEs face informational asymmetries and barriers to entry often controlled by large-enterprise monopolies. The government needs to do more to support SME innovation by removing these obstacles, such as the following measures.

Resolve Investment and Financing Problems for SMEs   One step is to establish a public fund to support venture investment. Several provinces and cities have already explored such funds, and their experiences can serve as a model on the national level. The national government should assess local experiences and also take into account existing tax policies at the national level. Such public funds would encourage venture investments and guide funds toward SME development. Another step is the possible creation of an SME credit guarantee scheme. SME guarantee agencies can be encouraged to implement risk compensation mechanisms. The government should also support the establishment of regional reinsurance agencies.

Establish Public Technology Service Platforms   The establishment and development of public technology service platforms are essential for China's long-term plans for innovation and development. There are several steps China needs to take in light of this effort. First, China needs to identify regions where technology service platforms deserve primary support from the government, based on regional economic and industrial characteristics. Second, performance criteria for monitoring and evaluation need to be developed. Third, the state should fund the purchase of R&D equipment for these platforms. Finally, these service platforms should be used to promote information diffusion and technology exchanges between enterprises and academics.

Establish a Sound Incentive Regime to Foster S&T Talents and Skills

High-level innovative talents and skills are needed for China to undertake more advanced scientific and technological research; thus, the country must develop a sound incentive system to promote their development. The objective is to foster a strong pool of highly innovative leaders in a variety of academic disciplines to form a community of Chinese talents for accelerating the pace of innovation. This can be done through incentive mechanisms and large national projects by assisting enterprises in attracting high-level talents and skills through measures such as offering stock options. Linking talent to performance and reward can also help to improve the quality of the workforce and raise the innovative capacity of Chinese firms.

Leverage Tax and Fiscal Policies to Promote Innovation

In recent years, public authorities implemented a series of policies designed to encourage indigenous innovation among enterprises. These include tax policies to encourage R&D investment, development of new technologies, and technology transfers. At the same time, special funds have been earmarked for supporting enterprises in the development of new products and technologies. In 2006, the Ministry of Finance participated in the formulation of several policies on scientific and technological development, including the "Outlined National Program for Medium- and Long-Term Development of Science and Technology (2006–2020)" (Ministry of Science and Technology 2006a) and the related matching policies for its implementation. Many of the policies stipulated are fiscal policies that have received positive responses from the private sector.

There are two main types of fiscal policies for innovation. One is direct public funding, and the other is tax-based fiscal incentives.

Direct Inputs of Public Funds   The direct use of public funds is still a major tool used by many countries to support indigenous innovation. One advantage is that it can pool funds for use in various activities, such as critical R&D or achievement of economies of scale that a fragmented market cannot accomplish. The United States, for example, has been relatively successful in using pooled funds for innovation. The public investments in R&D in the United States exhibit two main characteristics. First, resources tend to be focused on major undertakings, and priority is given to several critical fields. A large portion of funds goes to agencies such as the National Institutes of Health (NIH) and the National Aeronautics and Space Administration (NASA) for development in key fields of science and technology, such as medicine, space and aviation technologies, and information technology. Second, government funds are used in conjunction with market mechanisms to support enterprises, especially SMEs, in indigenous innovation. Thus, while public funds play a strong role in innovation policy, it is by no means the only source of funding. Additionally, public funds can lead to spillover effects that promote the use of private funds for R&D. One example of this is public investments in defense industries that generate spillover effects in related private industries.

Given the current level of indigenous innovation and the constraints faced by China, public finance should give priority to the common and core technologies areas as well as to the development of S&T infrastructure and a conducive environment for indigenous innovation such as that described above.

Tax-Based Fiscal Incentives   Fiscal policy—taxation in particular—makes full use of market mechanisms to offer incentives to firms to allocate more resources to R&D. Income tax deductions and accelerated depreciation are two tax policy tools that help firms increase their investment in R&D. Well-designed fiscal policies are therefore especially useful to address the lack of incentives for innovation, and good fiscal policies must take into account several considerations. Some examples of the key points follow.

First, income taxes should be uniform for both domestic enterprises and foreign invested enterprises. Additionally, geographical restrictions and tax differences for new high-tech enterprises should be eliminated. This is to ensure that these enterprises enjoy the same advantageous tax treatment even if they are not located within high-tech industrial parks, or special economic zones. It is important to establish a level playing field, which helps domestic companies and SMEs in particular to compete fairly and equally in the market.

Second, reforms of the value added tax (VAT) should be sped up to encourage equipment modernization and technological upgrades. VAT standardization will address problems in the tax system, such as double taxation.

Third, greater pre–income tax deductions should be offered for enterprise investments in R&D.

Fourth, China should also consider adopting universal import customs and import VAT deductions and exemptions for importing equipment used for R&D and testing purposes. This will help to standardize import taxes and encourage greater technology imports for indigenous innovation.

Fifth, China should permit enterprises to establish R&D funds with a percentage of sales revenues (for example, 3–5 percent) exclusively for the use of technological R&D.

Sixth, Enterprises should be permitted to accelerate the depreciation of R&D instruments and equipment.

Establish Government Procurement Practices to Support Enterprise Innovation

Government procurement will promote and encourage innovation in enterprises, as many surveys and practices observed. As such, reforms of government procurement practices should be considered to support innovation in Chinese enterprises.

First, establish first-buyer and ordering systems to offer incentives for indigenous innovation. Government can play an important role by being one of the first in the market to purchase new products developed by enterprises. In this way, government provides incentives for firms to innovate by reducing risks and ensuring that firms have a guaranteed customer base through government procurement.

Second, expand the scope of government procurement. Currently, government procurement is limited only to the purchase of goods and services by government agencies and public service institutions. Government procurement does not cover purchases by public enterprises, by the military, or for national construction projects. The State Council should coordinate government procurement policy, particularly for national construction projects, and the use of public funds for equipment and military purchases.

Third, direct social funds toward the purchase of innovative goods and services. To support the entry of innovative goods and services into the market, the government could establish a risk compensation fund for using innovative products to provide greater incentives for early adopters of new products. These incentives can be financial, such as partial rebates or refunds, offered to entities or private individuals for their first purchase of new innovative equipment or products. Another option is to offer compensation for firms and individuals for losses incurred in the use of new equipment and products. In both cases, the key goal is to increase incentives for early adoption and lower the risks associated with using new products and technologies.

References

Government of China. Official Documents. 2005. "Bulletin for the 5th Plenum of the 16th Central Committee of CPC." [In Chinese.] http://www.gov.cn.

Lv Zhen et al. 2005. *China Industrial Development Report*. [In Chinese.] Beijing: Economic Management Press.

Ministry of Science and Technology. 2005. "Statistics of Hi-Tech Industry 2005." [In Chinese.] http://www.most.gov.cn.

———. 2006a. "Outlined National Program for Medium- and Long-Term Development of Science and Technology (2006–2020)" [In Chinese.] http://www.gov.cn.

———. 2006b. "Statistical Results about Scientific and Technological Activities in Private S&T-Based Firms 2005." [In Chinese.] http://www.most.gov.cn.

National Bureau of Statistics. 2006. *China Statistical Almanac 2006*. [In Chinese and English.] Beijing: China Statistics Press.

7

China's Fiscal Policies for Innovation

Shahid Yusuf, Shuilin Wang, and Kaoru Nabeshima

Over the medium and longer term, China's growth rate, as well as the competitiveness of Chinese firms, will be increasingly linked to the country's technological capability. How this capability evolves from the stage of technology absorption and assimilation to one of independent innovation will depend upon the investment in research and development (R&D), in science and technology (S&T) workers, in information technology (IT), and in the development of several complementary institutions, which together feed into the making of a national innovation system.

Fiscal incentives impinge on the key determinants of innovation supply, including, in particular, R&D spending; hence, fiscal policy has a major role in the shaping of the innovation system. China already has deployed a wide range of fiscal and other incentives to promote technological capability. As this chapter shows, these are now beginning to yield results, which are reflected in rising R&D spending, patents, published scientific papers, increasing flows of S&T workers, and foreign direct investment (FDI) in R&D centers. These are all positive developments. As China moves forward, however, cross-country experience summarized in this chapter can usefully inform the necessary reforms in fiscal and other complementary measures for innovation.

The Case for Strengthening Innovation Capability

China has sustained a growth rate of over 9 percent for 25 years, in large part because, as in the Republic of Korea and Japan, a substantial slice of the huge investment of capital has effectively created world-class manufacturing capability. Key manufacturing subsectors are the drivers of the economy, and barring unforeseen changes in the demand for their products in foreign and domestic markets, they will continue to pull the economy forward over the medium term. Looking ahead, however, China's growth strategy is in need of adjustment for at least two reasons. First, the current approach is excessively reliant on vast inputs of capital and labor. China invests close to 45 percent of GDP, more than any other country in the world. While the composition of investment (which favors infrastructure, housing, and heavy industry) is partly responsible for this, allocative inefficiencies and the quality of the investment are contributing factors.

Sources of Growth

China's economic expansion is costly in terms of capital expended and, relatedly, of the consumption deferred. Energy inputs per unit of GDP and material inputs per unit of industrial output also point in the same direction: the system is delivering growth but the economy is paying a substantial price. There is considerable room for raising efficiency and technological capability so as to squeeze more value added from resource inputs.

The industrial sector, which is the principal economic engine, would also benefit from gains in both these areas. Currently, the global competitiveness of China's leading manufacturing sectors rests upon low input costs, scale of production, technology absorption, speed of response to market demands, and the fulfillment of orders, and increased attention to the quality of products. Particularly notable are the development of local skills in organizing and managing complex production tasks, the harnessing of imported technologies, and the acquired ability to respond quickly to the needs of buyers from throughout the domestic market and across the world.

Through a continuing refinement of these skills, coupled with the elastic supplies of both capital and labor, the Chinese growth model should deliver good results for several years. However, both the returns on investment over the medium run as well as the economy's longer-term growth potential will increasingly depend upon the nurturing of a broad-based innovation capability. Innovation alone will not raise economic efficiency, increase productivity, or enhance the competitiveness of the tradable sectors, but cross-country experience suggests that its contribution will be substantial. In fact, the competitiveness of manufactured products, producer services, and the creative industries is becoming ever more closely tied to incremental process innovation, product innovation, and innovations in design, organization, and logistics. Building a broad-spectrum innovation system is now widely viewed as the central plank of a growth strategy for industrialized and industrializing countries alike (see Steil, Victor, and Nelson 2002).

Such a system is made more urgent by the strengthening demand for innovation, which is the result of the global integration of product and labor markets, the codification of technologies that is accelerating diffusion, the associated commodification of many products and services, and the need to ameliorate the scarcity of certain raw materials as well as environmental externalities.

Increased Innovation

While innovation policies must address many factors that impinge upon the incentives to innovate and the demand for innovation (such as entry barriers for start-up firms, protection of intellectual property, and the availability of venture capital), the main thrust of government technology policy is directed toward the above four determinants of innovation supply—measures to increase research and development, technology transfer, S&T manpower, and the use of information technology—are sound reasons for such emphasis.

First, a wealth of empirical evidence indicates that private returns from investment in research and development average 28 percent (Wieser 2005). Moreover, because the spillover benefits within and among industries are large, the social rates of return can be as high as 90–100 percent. The public good characteristics of scientific knowledge, the massive divergence between high private returns and much higher public returns, and the comparability of returns among countries make research and development a worthy target for public policy (see also Hu and Mathews 2005).

Second, the links between R&D spending and the increase in productivity have been rigorously established. Studies on countries in the Organisation for Economic Co-operation and Development (OECD) show that the elasticities of total factor productivity with respect to R&D lie in the 0.03–0.38 range. They are higher in the United States than in Europe and Japan, pointing to differences in the efficiency with which R&D resources are utilized and the nature of enabling institutions (Wieser 2005).

Third, S&T manpower complements R&D spending and affects the returns from this outlay. If technological skills of the requisite quality are not forthcoming, expenditures on research yield small benefits. Moreover, the relationship running from human capital to productivity has been established through numerous studies (see for instance Nabeshima 2004; Ciccone and Papaioannou 2005; Temple 1999, 2001).

Fourth, for late-developing countries such as China, the initial catch-up stage is dependent upon technology absorption and assimilation from abroad. This can be expedited through heavier investment in imported plants and equipment that embody the latest technologies—which Japan did in the 1950s; through licensing of foreign technologies or reverse engineering—the path taken by Korea; or by attracting foreign direct investment into medium and high-tech sectors and requiring investors to share technology—which is the strategy adopted by Japan, China, and particularly Singapore. Borrowing from the global technology bank is a low-cost way of scaling the technology ladder, and in many subsectors this is the shortest route for China to the frontiers of technology (see Hu and Mathews 2005).

Fifth, and finally, the revolution in IT has added a new determinant of productivity. Recent research, mainly on the OECD countries, shows that the main IT-producing and IT-using sectors were the leading contributors to the gains in productivity achieved since the mid-1990s (Gordon 2004; Jorgenson and Motohashi 2005; OECD 2004). The rapid growth of the electronics, computer-producing, and telecommunications subsectors has significantly boosted productivity in manufacturing. But arguably, even more important are the big advances registered by financial services, retailing, wholesaling, and logistics, all of which have been quick to exploit the possibilities opened up by IT. Aside from the incorporation of IT hardware and digitization of data, the major benefits to the services industries have been achieved by new business models, efficient processing of huge amounts of information, and an overhaul of the organization of work. The Wal-Mart effect is synonymous with the surge of productivity in the United States (*McKinsey Quarterly* 2002; Jorgenson et al. 2007).

Policy Levers and Policy Directions

In a market economy or in the Chinese context, the demand for innovation or incentive to innovate can strongly stimulate R&D spending, technology transfer, an increase in the supply of S&T skills, and the greater use of information technology. Everywhere, this is seen happening as a result of the apparently accelerating tempo of technological change and globalization. But governments worry that because of the public good aspects of knowledge and because of market imperfections, not enough resources are devoted to accumulating knowledge, resulting in an undersupply of innovation and suboptimal growth of productivity. Governments also worry that if other countries invest more in research and development, there is the danger of being left behind. The desire to remain at the leading edge of technological development or to move to the technological forefront in selected areas has spurred a "technological arms race" and concentrated attention on rankings of technological capability and preparedness. From being a minor strand of growth policy, the effort to strengthen the innovation system is now at the very center and the focus of several intersecting policies. Among these, a number of fiscal (tax, expenditure, and subsidy) policies loom large.

The purpose of these policies is to do the following:

- Increase the overall level of R&D spending throughout the economy.
- Affect the distribution of R&D among sectors so as to favor the more promising sectors and loosen constraints, such as the supply of S&T workers. Most of the R&D outlay across countries is in IT hardware, automotive industries, and pharmaceuticals, although R&D intensity is greatest in software and IT services, pharmaceuticals, and IT hardware (see van Pottelsberghe de la Potterie 2008).
- Directly support certain technology development in activities assumed to have high long-run social returns through direct budgetary grants and subsidies.
- Encourage capital investment in imported plant and equipment in part to promote technology absorption.
- Draw FDI into selected industries to serve as a conduit for new technologies and possibly also as the axes for cluster development in urban technology zones and parks.

Recent Trends in R&D Policies

To achieve these ends, governments—national as well as subnational—use a small number of taxes, subsidies, grant mechanisms, and financing techniques. These actions need to be coordinated with other policies and institutions, and European as well as East Asian economies, which are building robust innovation systems, are the ones giving closest attention to such complementarities. Over time, the trends among the OECD countries have been as follows:

- Away from direct grants for R&D outlay, whether for research in general or R&D in specific areas and toward the greater use of tax credits for R&D spending. This allows firms and market participants to select the most promising fields to invest in. Many governments offer more generous fiscal incentives to small and medium enterprises (SMEs) to encourage new starts and the growth of small firms.
- With the lion's share of the R&D, generally two-thirds or more, done by the corporate sector labs (including in China), followed by universities and private research institutes, with public research institutes generally coming last, although this is not uniformly the case.
- A reduction of grants for defense-related research conducted by public or private entities.
- A reduction in budgetary support for public research institutes and either a stabilization of or a decline in fiscal transfers to publicly owned universities.
- A gradual reduction in publicly provided venture capital for SMEs and technology-intensive start-ups, although bodies such as the Small Business Innovation Research program in the United States remain important and have been widely imitated. Likewise, venture capitalists supported through budgetary grants are significant sources of financing in many countries, including Singapore and Korea.

The above trends serve to offset distortions in the financial system and to replace so-called missing markets. Of the US$677 billion spent on R&D globally, more than half was done by multinational corporations, and in many instances their expenditure exceeded that of countries such as Spain and Belgium. The important characteristic of multinational corporations'

R&D is its increasingly international character. For example, 13 percent of the U.S. multinational corporations' R&D and 43 percent of Swedish multinational corporations' R&D is conducted by their foreign affiliates. Worldwide, this percentage is on the rise, with 16 percent of R&D conducted by foreign affiliates of multinational corporations, compared with 10 percent in 1993. East Asian economies, and China especially, are benefiting disproportionately from the internationalization of research (see figures 7.1–7.3). The attractiveness of East Asian economies to multinational corporations is dependent on the supply of S&T workers, the nature of the overall innovation system, and the quality of the information and communication technology infrastructure that facilitates a dispersal of research (UNCTAD 2005).

Central government fiscal incentives for FDI, the creation of R&D facilities by multinational corporations, and investment in science parks (where provided) have stabilized or declined; however, subnational governments continue to compete for FDI by offering fiscal, financial, or infrastructure-related inducements, and they continue to invest in science parks, sometimes with the help of central government fiscal support. The research on the sensitivity of FDI to tax incentives is mixed, and as a recent World Bank study shows, many countries have failed to attract FDI in spite of fiscal inducements (Zhao 2005). But few governments are willing to risk abolishing the tax holidays offered. Similarly, although a recent cost-benefit analysis of science parks indicated that they have little regional developmental impact, many governments in industrializing countries continue to provide fiscal support or financing through other channels (Wallsten 2004). New parks are sprouting throughout Asia, although it is unclear whether the taxpayer gains the benefit of high social returns.

Fiscal Instruments and Effects

A small number of fiscal measures have been deployed to promote technology development in the OECD countries (see table 7.1). Among these are five leading tax instruments:

- R&D tax credits
- Capital gains tax
- Investment tax credit
- Structure of direct taxes and loss carry-forward provisions
- State-level incentives.

Figure 7.1 *R&D Expenditures of Selected Multinational Corporations and Economies, 2002*

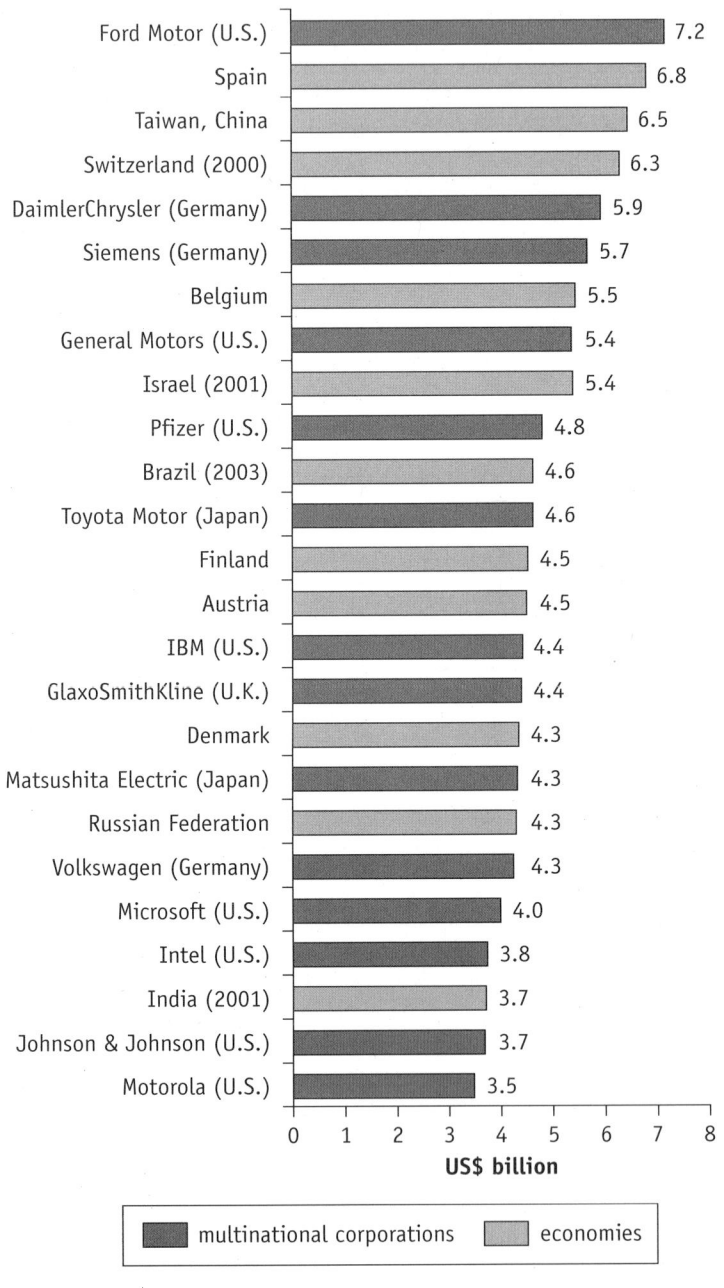

Source: UNCTAD 2005, figure IV.1.

Figure 7.2 *Locations of R&D Investments by Multinational Corporations, 2004* (percentage of multinational corporation responses)

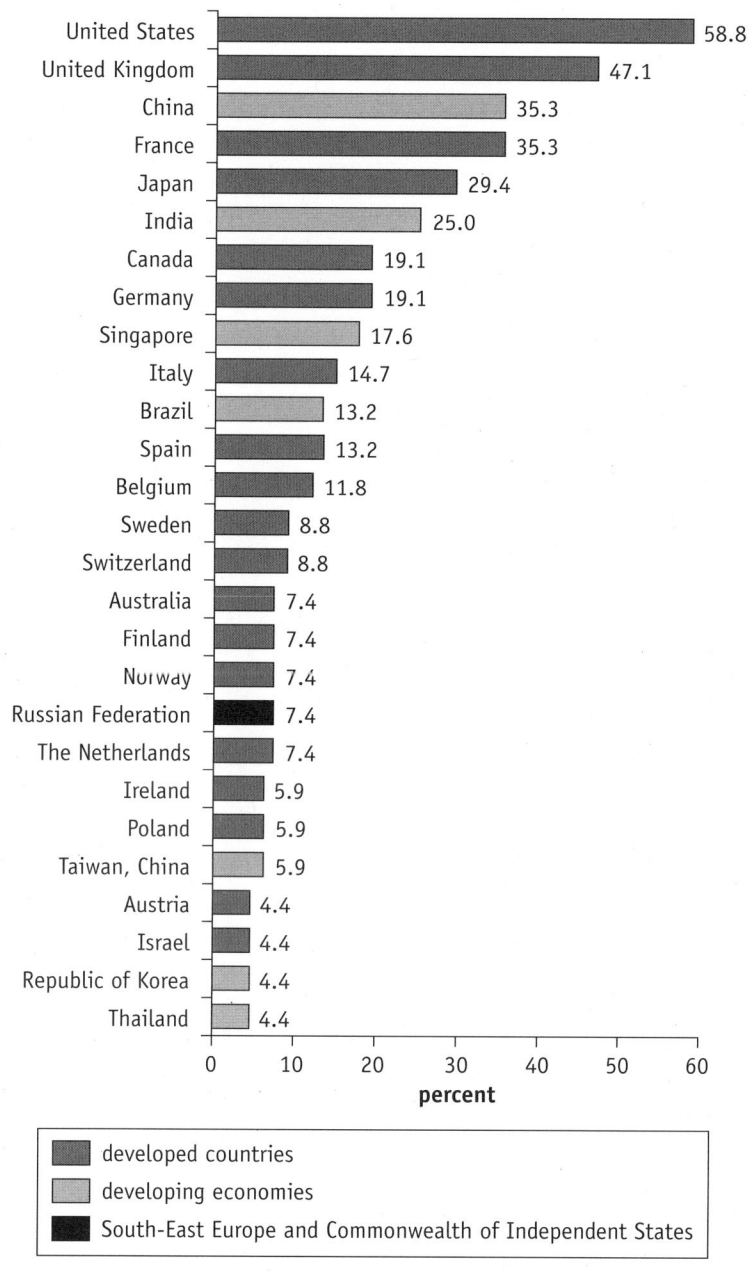

Source: UNCTAD 2005, figure IV.8.

Figure 7.3 *Prospective R&D Locations Considered Most Attractive, 2005–09* (percentage of multinational corporation responses)

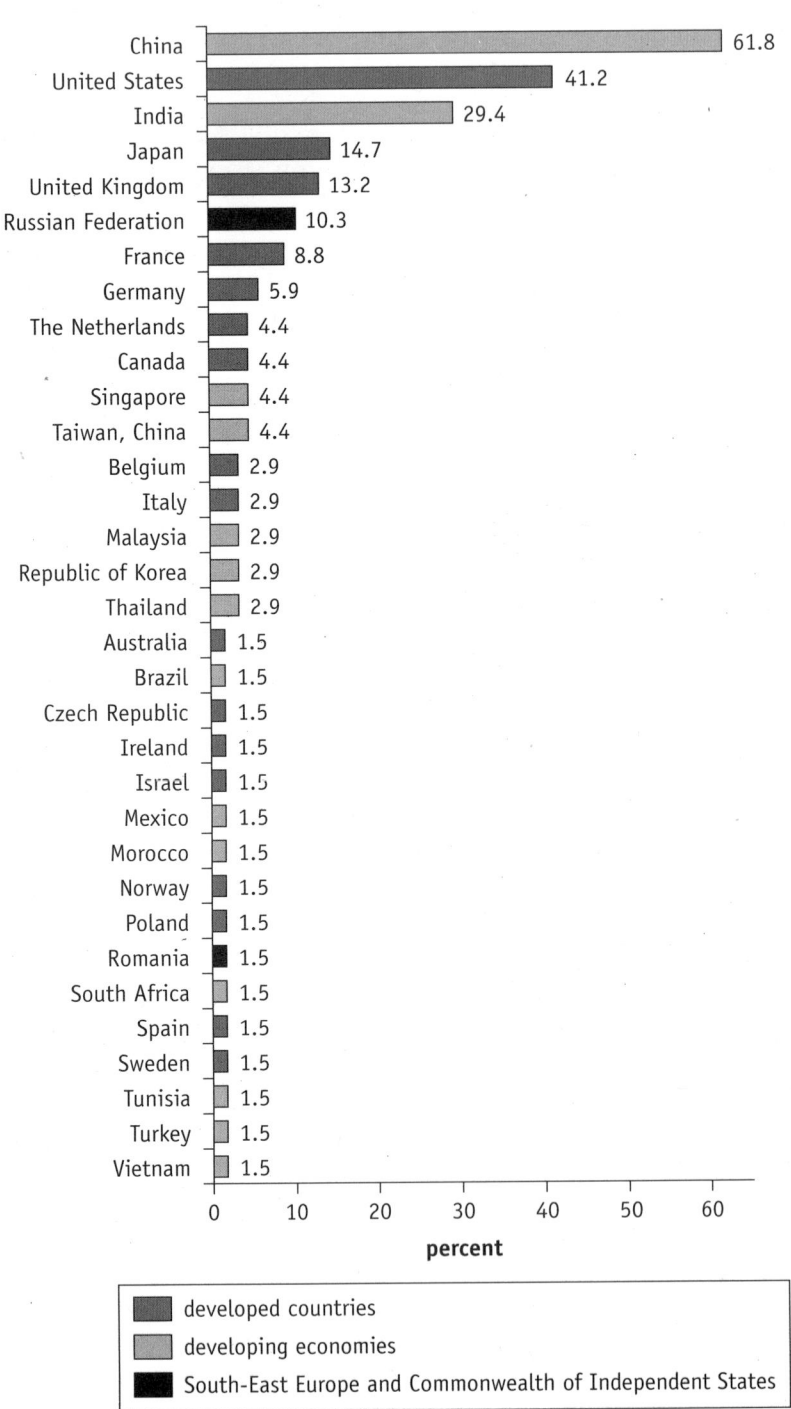

Table 7.1 *Various Types of Fiscal Incentives, Selected European Countries*

Country	Type	Main instruments	Main feature
Finland	Selective project funding	Selective project funding	Funding criteria redefined to meet needs of new and small companies; flexible and interactive funding process to ensure easy access
France	Incremental tax incentive	Credit Import Recherche (CIR)	Reduction in tax equal to half of R&D expenditure over one year minus average expenditure of the same type over the two previous years with a limit of €6.1 million per company per year
Netherlands	Tax credit	Wage Tax and Social Insurance Act	R&D wage cost reduction: 40% over the first €90,000, 13% over the rest; for start-ups: 70% over the first €90,000; maximum €8 million
Spain	Mix direct and indirect incentives	National Plan for Scientific Research and Technological Innovation 2000–03	Financial incentives (subsidies, refundable loans, 0% interest rate); tax credit on R&D expenditure (mixed-regime incremental and volume-based, rate ranging from 30% to 10%); special measures for SMEs (favorable tax regime)
United Kingdom	Volume-based tax credit	SME tax credit; large company tax credit	50% super-tax deduction, rebate if not in profit
Norway	Volume-based tax incentive, with a cap	n.a.	25% deduction for all R&D expenditure up to a limit of NKr 4 million; expenditures buying R&D
Portugal	Mixed tax credit scheme	n.a.	Basic deduction rate of 8% in 1997–2000; 20% from 2001; incremental deduction rate on R&D expenditure over the average of the previous two years

Sources: Cherbonnier 2002; Hervik 2002; Hezemans 2002; Kemppainen 2002; Pedrotti 2002; Rosa 2002; Santos 2002.
Note: € = euro; n.a. = not applicable; NKr = Norwegian kroner; R&D = research and development; SME = small and medium enterprise.

On the R&D tax credit, the voluminous research is essentially positive: the credit does stimulate R&D and increases welfare. However, the gains derived depend on the design of the instrument, for example, volume-based or incremental (which affects deadweight losses); temporary or permanent; simplicity; whether it is subject to a cap; the administrative burden imposed on companies by the claims process; and the speed with which firms are reimbursed. A volume-based tax with no cap, with extra provisions for SMEs, that is permanent and simple to claim is the most attractive, although it does entail higher deadweight losses (see Bloom, Griffith, and van Reenen 2002; Hall and van Reenen 1999; Russo 2004).

For many new starts and especially high-tech firms, the greatest attraction for entrepreneurs and venture capitalists is the prospect of "going public" or being bought by other firms and disposing of the company's shares at favorable prices. The reward for risk taking is greater if the capital gains tax is low, as is the case in the United States. This explains in part the number of new starts in certain subsectors such as IT and biotechnology, which have a disproportionately large role in developing new technologies and in commercializing innovations. These firms, catalyzed by the venture capital sector, are one reason why the U.S. innovation system is such an efficient user of R&D funding.

The structure of the corporate and income tax rates and the loss carry-forward provisions also provide incentives for entrepreneurial risk-taking behavior, which is critical for the transition from research findings to innovation and its marketing via a product or a service. Low corporate rates encourage incorporation of businesses, while relatively higher income taxes and loss carry-forward permit a bigger sharing of the risks with the state. The option of incorporating and being taxed at a lower marginal corporate rate subsidizes risk taking and, again, is an inducement to the formation of new businesses (Cullen and Gordon 2002; Russo 2004).

The investment tax credit can strongly incentivize spending on plant and equipment, depending of course on the terms. Because so many of the early gains to industrializing countries accrue through technology, which is embodied in new equipment, and because higher levels of investment are also associated with "learning by doing," tax measures that raise the level of industrial investment can, within a competitive market environment, increase technological assimilation and productivity growth. China itself is deriving benefits from the sustained investment in the manufacturing sector, but in some sectors a tendency to overinvest could be tackled through tax reforms, including the phasing in of the consumption VAT.

In federal or decentralized fiscal systems, state- or province-level taxes introduce additional incentives or disincentives to R&D and innovative activities. In the United States and also in China, these subnational incentives play a considerable role, though possibly not in determining the overall level of R&D as much as in influencing its geographic location.

Complementary Policies

Fiscal instruments are an important category of measures for building the innovation system. They must necessarily be complemented by several others to reinforce and possibly magnify the effects of tax policy. Among the several candidates, two merit special attention. In the experience of OECD and East Asian economies, there are many instances where direct government research support to companies or research institutes or universities was vital for the conduct of research—basic or applied—and for development of new technologies (Hu and Mathews 2005). In fact, close to 10 percent of all private sector R&D in OECD countries between 1991 and 2001 was funded by the government. Research supported by the U.S. government has been critical to technological advances in the aeronautic, biotechnology, and pharmaceutical industries and to the creation of the Internet. In Israel, government investment has been responsible for the strides made by the IT sector; in Korea and Taiwan, China, publicly financed R&D provided the technological foundations for the electronics industry (Hu and Mathews 2005), as it did also for the electronics and computer industries in Japan.

Policy makers in East Asia, as well as their counterparts in the West, view the provision of government research grants for specific sectors or projects as desirable investments in research capacity building and knowledge acquisition. Certain sectors, such as IT, automotives, and pharmaceuticals, attract a disproportionate amount of R&D, private and public. Such strategic technology policy interventions (for example, in biotechnology, new materials, and nanotechnology; see Jaffe, Lerner, and Stern 2002) are a means of circumventing perceived market myopia or market failures. Because private returns diverge from social returns—especially so for speculative and risky research ventures—without a reasonably secure medium-term payoff, private firms shy away from putting money into these activities. They are particularly averse to investing in basic research, which approximates a pure public good and often is perceived as having little discernible commercial outcome.

In the United States, the National Science Foundation and the Defense Advanced Research Projects Agency, among several other agencies, finance such "blue skies" research, betting on the likelihood that it will generate high social returns and, possibly, significant private returns down the road. Many research endeavors prove to be dry holes, but strategic technology policies are buoyed by the spectacular successes such as the Internet and the many commercial products that can be traced back to esoteric defense research projects (see Hambling 2005).

Blue skies basic research and applied research in promising areas will require direct grant funding because tax-based inducements generally might not work; also, because of missing financial markets for certain kinds of quasi-public goals, the private funding from venture capitalists or others might not be forthcoming. There is no reliable rule of thumb as to how much of the budget should be allocated for this type of research. However, late starters with a lot of catching up to do and much knowledge to assimilate may need to set fewer funds aside for this purpose, although the need to build research capacity for the future increasingly argues for greater attention to basic research than has been the norm to date.

Policy makers worry, and rightly, that direct grant funding for R&D, whether to private entities or to public research institutes, could displace private resources, which would have been forthcoming. Much empirical effort has gone into assessing the degree to which public support for R&D substitutes for or complements private R&D. The existence of complementarity is weakly supported by a review of some 30 studies (see David, Hall, and Toole 2000), but a meta-analysis of their research yielded inconclusive results (García-Quevedo 2004), and there are findings that suggest that direct public spending on R&D (as distinct from tax incentives for the private sector) crowds out private R&D (Mamuneas and Nadiri 1996).

Government support for R&D must also take account of two other factors: one is the supply of S&T workers; the other and related factor is the economy's absorptive capacity for additional R&D spending. If the supply of S&T workers is inelastic, then pouring more resources into R&D will only raise the wages of existing workers and will have limited impact on the volume of research conducted. Similarly, if the number of experienced research scientists and managers is limited and cannot be readily expanded, or the organizational capacity for handling research is underdeveloped, or the physical infrastructure of labs and support facilities is sparse and cannot be expanded rapidly, then again the returns to an increase in R&D spending will be small. Some countries such as the United States, the United Kingdom, Canada, Singapore, and others have managed more or less successfully to sidestep the constraint by importing and absorbing foreign S&T workers. Korea and Taiwan, China, relied on the reverse flow of their citizens who had trained and settled mainly in the United States. However, this source of supply has its limits. It does not easily provide the desired mix of skills, and it generally does not bring in the experienced researchers, team leaders, and research managers whose presence frequently determines the pace and productivity of R&D.

In other words, government plans to raise R&D spending through grants and tax incentives and inducements for FDI must be calibrated with reference to complementary budgetary and institutional resources that will increase the supply of S&T workers (from domestic and foreign sources), bring in the appropriate mix of workers, and also put in place the R&D infrastructure, hard as well as soft. This can be a slow process, but unless it is tackled, the incremental R&D spending is unlikely to lead to greater innovativeness or productivity.

Absorptive capacity for R&D and the environment nurturing creativity cannot be expanded overnight—even a decade might be too short a time. Building technological capability is a necessary objective for a rapidly industrializing country. Technological capability is a function not just of R&D spending and the total number of S&T workers; it also is a function of experience, scientific traditions, social capital, professional networks, openness to ideas, tacit knowledge of researchers, and multiple interlaced institutions that sustain the research community and give researchers the space, the freedom, and the intellectual property rights to do good research (Scotchmer 2004; Skinner and Staiger 2005).

While R&D spending, patents, and published papers are convenient indicators for policy makers, it is the quality of the research that impinges on the bottom line of productivity. The vast number of patents registered remain "dark patents," never cited and with no measurable scientific or commercial outcomes. Likewise, no more than 5–10 percent of the science literature—mainly that published in the top journals—has an influence and contributes to the conversation on scientific issues; the rest vanishes without a trace (van Dalen and Klamer 2005). Building a strong innovation system should be a priority for China, but the experience of more advanced countries argues for "making haste slowly."

China's Innovation Position in Global Context

Strengthening the innovation system is rightly one of the main objectives of China's new 11th Five-Year Plan. China's R&D spending increased from 34.8 billion yuan (Y) a year in 1995 to about Y 300.3 billion in 2006. As a share of GDP, R&D spending increased from 0.6 percent in the mid-1990s to 1.4 percent in 2006 (see table 7.2). This share is the highest in the developing world, higher than India's and Brazil's but still lower than the world average (1.6 percent) and that of developed countries (2.2 percent). The share of government expenditures has remained at about 4 percent since 1998. China now ranks sixth in the world in terms of its total R&D expenditure (second, if a purchasing power parity, or PPP, exchange rate is used). However, it should be noted that China's expenditure statistics are not comparable to those for OECD countries because China does not attach values to tax incentives (expenditures) whereas OECD countries do.

From the breakdown of R&D expenditure, it is apparent that large and medium Chinese firms have become the main investors in R&D activities. For example, in 2006 China's R&D spending totaled Y 300 billion, of which 25 percent came from government allocation and 70 percent was from enterprises' investment (see table 7.3).

Large and medium enterprises are now becoming big players in both financing and conducting R&D programs in China. According to a report on firms' input into R&D activities, issued by the Central State Assets Supervision and Administration Commission in August 2005, the total R&D

Table 7.2 *China's R&D Expenditures and Shares in Relation to GDP*

Item	1995	2000	2006
GERD (billion yuan)	35.0	89.6	300.3
Growth (percent)	—	16.9	22.6
As percentage of GDP	0.6	1.0	1.4
Govt. S&T appropriations (billion yuan)	—	57.6	136.8
As percentage of govt. expenditure	—	3.6	3.4

Source: National Bureau of Statistics 2007.
Note: GERD = gross domestic expenditure on research and development; S&T = science and technology; — = not available.

Table 7.3 *Distribution of Funding on R&D, 2002–06*
(percent)

Item	2002	2003	2004	2005	2006
R&D institutions	27.3	25.9	22.0	20.9	18.9
Universities	10.1	10.5	10.2	9.9	9.2
Enterprises	43.5	46.8	48.5	51.0	54.3
Others	19.1	16.7	19.3	18.1	17.6

Source: National Bureau of Statistics 2007.

input in 2004 by state-owned enterprises, which it supervised, came to Y 76.8 billion, accounting for 1.5 percent of firms' annual sales revenues. The share for all state-owned industrial enterprises was even higher (2 percent).

Basic research accounts for about 5 percent of total R&D spending, applied research for another 20 percent, and experimental and developmental research account for the balance (see table 7.4). Universities' spending

Table 7.4 *Distribution of R&D Expenditure, 2002–06*
(percent)

Item	2002	2003	2004	2005	2006
Basic research	5.7	5.7	6.0	5.4	5.2
Applied research	19.2	20.2	20.4	17.7	16.8
Experimentation and development	75.1	74.1	73.7	77.0	78.0

Source: National Bureau of Statistics 2007.
Note: Basic research means the experimental or theoretical research intended to obtain new knowledge on basic principles of phenomena and observable evidence (revealing the essence, identifying laws of operation, and getting new development and theories), not aimed for any specific purpose or usage. *Applied research* means innovative research intended to define the possible application of basic research results, or to find out the new methods (principle based) or new approaches for achieving predefined targets. Applied research is mainly aimed for a specific purpose or objective. *Experimental and developmental research* describe systemic work intended to develop new products, materials, and equipment, for setting up new processes of production, systems, and services, as well as essentially renovating those above-mentioned works, by using existing knowledge obtained from basic research, applied research, and empirical experiences. Definitions are from *China Science and Technology Statistics* (2003): R&D Expenditure. http://www.sts.org.cn/tjbg/zhqk/documents/2003/1028.htm.

on R&D is about 10 percent of the nation's total, and over 50 percent of their funding comes from government appropriations, while funding from enterprises accounts for one-third. Universities' expenditure on R&D has increased rapidly since 2000, at an annual rate averaging over 30 percent.

Overview of Current Fiscal Incentives Given to Stimulate Spending in Innovation

China is using various fiscal incentives to enhance R&D and technological innovations, such as tax holidays, preferential tax rates, accelerated depreciation, import duty exemption, export subsidies, reduced tax rates, and tax rebates. Detailed explanations of each policy are given below.

Fiscal Incentives for R&D and Related Activities

The Chinese government provides import tariff exemption in the following cases:

- To facilitate firms' technological renovation and product upgrading in existing state-owned enterprises. In addition, targeted industries such as those in the electronics sector were exempted from tariffs and import-related VAT on equipment during the 9th and 10th Five-Year periods.
- To promote technical transfer and commercialization. Foreign individuals, firms, R&D centers engaged in activities of consulting, and technical services related to technology transfer and technological development are exempted from corporate tax on their incomes.

Fiscal Incentives Given to Various Technology Development Zones

Establishing economic zones, new and high-tech industrial zones (HTIZs), and economic and technological development zones is one of the key measures the Chinese government has adopted in facilitating acquisition of new and advanced technologies, promoting technological innovation, promoting the commercialization of S&T results, and enhancing China's industrial competitiveness. From the early 1980s, China began establishing special economic zones and, since the 1990s, high-tech industrial development zones.

In 1991 China approved 21 national HTIZs, and by 2005 the total number countrywide had risen to 150, of which 53 are at the national level.

These HTIZs have nursed 39,000 high-tech firms employing 4.5 million people. The total turnover of firms reached Y 2.7 trillion in 2004, an increase of 31 percent over the previous year. The per capita profit was Y 33,000; per capita tax yield was Y 29,000, and the per capita foreign earnings were Y 157,320.

In the national HTIZs a series of investor-friendly policies and measures have been introduced. These measures include tax reduction and exemption policies.[1]

Fiscal Incentives Related to Income Tax

The Chinese government offers various tax holiday schemes to different types of firms.

Income tax

- Foreign-invested enterprises can enjoy the preferential treatment of income tax exemption in the first two years after making profits and an income tax reduction (by half) in the following three years.
- Foreign-invested high-tech enterprises can enjoy income tax exemption in the first two years after making profits and an income tax reduction (by half) in the following six years.
- Sino-foreign joint ventures can enjoy income tax exemption in the first two years after making profits.
- Other firms are eligible for income tax exemption in the first two years when starting productive operation.
- Domestic firms in HTIZs are eligible for preferential treatment but with limits in terms of types of business activities (income earned from technology transfer or activities related to technology transfer, such as technical consulting service and training). A ceiling is imposed on how much they can benefit from income tax exemption (less than Y 300, 000).
- Income tax rate is set at 15 percent in these zones, which is much lower compared with the normal rate for those located outside the zones. Firms whose export share is above 70 percent of their annual production can enjoy further income tax reduction (10 percent).

Turnover tax

- Foreign enterprises and foreign-invested enterprises are also exempted from the business tax on technology transfer.

Tariff and import duties

- Tariff and import-stage VAT exemptions have been granted to foreign-funded enterprises for their importation of equipment and technologies that are listed in the "Catalogue of Encouragement" issued by the Ministry of Commerce, National Development and Reform Commission, and China State Tax Administration (2003).

Accelerated depreciation

- New and high-tech firms are granted accelerated depreciation for equipment and instruments (since 1991; see China's State Council Document [1991], No. 12).

Direct Grants Given for Specific R&D Activities by the Central and Subnational Governments

China has used direct grants through government budgetary allocations or appropriations to promote scientific and technological development for decades. Currently, the funding sources are diversified, and the government is no longer the only financier for new technology development. But many key scientific and technology projects are still financed by the state, particularly the central government.

China carried out a series of programs to support R&D activities and S&T development. During the sixth Five-Year Plan period, China entered into 3,000 contracts to import high-tech and new technologies from developed countries. Projects that have received large direct grants from the central government since the 1980s include the following:

- National high-tech R&D programs (863 programs)
- National Key Technologies R&D Program
- Torch Program
- National Program on Key Basic Research Projects
- Small and Medium-Sized Enterprise Fund
- National key laboratories (initiated in 1984). China's national key laboratories now total 183 and are located mostly in universities and in the China Academy of Science.
- The "973" program, established in 1999 to support basic science research and innovation. The key research areas include agriculture, energy, information, resources and environment, population and health, new materials, across-discipline research, and more.
- The China Technical Innovation Fund, set up in 1998.

In addition, China also created industrial funds to promote the development of targeted industries. For instance, a fund was established during 1986–2001 for electronics and information technology, and a total amount of Y 2,456 million was invested in domestic firms. Starting in 1999, China initiated programs to finance the research on biotechnology products with 140 projects funded by a total of Y 16 billion. Subnational governments also sponsored many similar programs or projects related to commercialization of R&D results, technical incubation, and others.

Scholarships for Students Studying in Science and Engineering Fields in China and Abroad

Training of Chinese students overseas is an important policy implemented by the Chinese government since the late 1970s to increase the supply of S&T skills. By 2003, China had sent over 700,000 students to study in 108 countries. From this total, over 527,000 persons are continuing their studies in foreign universities, conducting research, teaching, or working overseas, providing a potential pool of skills for China to draw on.

The Chinese government has created an Overseas Study Fund to sponsor Chinese students and scholars to pursue their studies or training overseas. In 2004, the fund sponsored 3,630 people for advanced studies or research programs overseas. In line with China's development priorities, the fund identified seven disciplines or academic fields as its sponsorship priorities for 2004:

- Telecommunications and information technology
- Agricultural science
- Life science and population health
- Material science and new materials
- Energy and environment
- Engineering science
- Applied social science and subjects related to WTO issue.

Incentives Given to Attract Overseas Chinese Back

In line with the efforts under way to build the national innovation system, China has adopted various policies to attract overseas Chinese back. The government sponsors Chinese scholars working in overseas universities or research institutions to return for scientific research or academic exchange positions in China, or work for up to one year in areas beyond their selected fields. These activities include, among others, joining the research programs sponsored by the state, ministries, or provincial governments; helping domestic institutions solve key scientific issues; giving lectures and conducting training; attending international conferences or important national meetings; and assisting in technology transfer and technical exchanges. The sponsorship consists of international travel and living allowances. The *Chunhui* program has sponsored 8,000 Chinese scholars with PhDs obtained overseas to come back to carry out short-term work. The Yangtze River Fellowship program awarded 537 overseas Chinese scholars professional appointments in Chinese universities for curriculum building and teaching and for joint academic research.

The government also sponsors Chinese students who have completed their studies overseas with a master's degree or above to return and work in scientific and technical sectors. Seed money is provided to enable them to start up their business. The applicants must meet certain criteria in order to get the state grant, such as a minimum of one year of study in a foreign higher-education institution, with a master's or PhD, acquired ability to conduct independent research work, or work on state-proposed projects. The grant can be as large as Y 200,000. The Ministry of Education set up a special fund to help returned overseas students initiate start-ups. The fund was set up in 1990 and more than 10,000 people have benefited from total funding of Y 350 million.[2]

China also established overseas student business bases or industrial parks jointly organized by the central and provincial governments to help returned overseas students initiate their businesses in China. The central and provincial governments share expenses for building infrastructure and other facilities and for providing services needed for starting up a new firm.[3]

Fiscal Incentives Given to Venture Capital Firms

The government's "Views on Mechanisms for Establishing Venture Capital and Investment," promulgated in 1999, initiated the development of the venture capital industry. By 2007, China had established close to 300 venture capital firms, with a total management portfolio of Y 50 billion invested in more than 2,000 projects. Close to 80 percent of the venture capital was invested in high- and new-tech projects; 85 percent of funded projects have a history of less than five years. Firms that are in the seed money stage or the growing stage account for 65 percent of the recipients. Funded firms with actual equity capital of less than Y 30 million account for 80 percent. Most of this venture capital goes into small and medium high-tech firms such as ICT and biotech.

Currently, China grants preferential tax treatment for foreign investment and venture capital firms. If venture capital firms are foreign owned or a foreign-Chinese joint venture (in which foreign investment is no less than 25 percent of total firm equity), they enjoy all the fiscal preferential treatment that is given to foreign funded enterprises, such as income tax exemption in the first two years after making profits and income tax reduction by half in the following three years. The income tax paid by foreign funded firms is 15 percent in economic zones, new and high-tech industrial zones, and economic and technological development zones, while domestic firms pay a 33 percent tax. In this regard, Chinese domestic investment and venture capital firms are at a disadvantage. The Chinese government introduced a new tax policy to reduce the effective tax rates to 17.5 percent for qualified venture capital firms.

Fiscal Incentives Given to Attract the Establishment of R&D Centers by Multinational Corporations

China has been recently successful in attracting multinational corporations to establish R&D facilities in China. These multinational companies are coming to China to take advantage of both China's human resource potential and the low cost of carrying out R&D in China. In Beijing and Shanghai, many multinational corporations have set up their offshore research facilities. According to the Ministry of Commerce, there were about 690 research facilities or research centers created by multinational corporations in China, and accumulated investment exceeded US$4 billion. For instance, Microsoft, IBM, Siemens, Motorola, Nokia, Samsung, and Dell have opened their China headquarters or regional research centers in Beijing. Microsoft built its first research institute in Asia in Zhongguancun district in Beijing in 1998. Intel also established its Asian research institute in Zhongguancun in 2004.

The abundance of academic think tanks, universities, and various research and development centers in Beijing and Shanghai is a major attraction for these multinational firms. In addition, the central government and municipal governments also offer incentives to attract these multinational corporations and induce them to set up their R&D centers. The fiscal incentives offered include the following:

- Exemption from import duties and import-related VAT for imports of equipment, devices, and spare parts for R&D purposes (1997).
- Tariff and import-related VAT exemption for acquiring imported new and advanced technologies. Foreign-funded R&D centers receive the same fiscal benefits as foreign-funded high-tech firms and enjoy the same fiscal preferential treatments (November 2004).
- Exemption from corporate tax for revenue earned through the delivery of consulting or other technical services related to technology transfer, and technical development activities (1999).
- Reduction in income tax payment for those R&D centers whose expenditures on R&D increased more than 10 percent annually.

Fiscal Incentives Given to Attract Foreign High-Tech Industries

Since the adoption of the policy of reform and the opening up to the outside world in the late 1970s, China has offered many incentives to attract FDI in high-tech industries (for example, tax holidays, preferential tax rates, import duty exemption, accelerated depreciation, and tax credits). In order to direct the FDI in compliance with the national industrial planning, China issued an interim regulation on FDI directions and issued the industrial catalog guiding foreign investment in June 1995. This regulation was revised later in 1997, and the revised text highlighted priority industries alongside the principles for complying with structural adjustment, the introduction of advanced technology, and encouragement to invest in the central and western areas.

In 2003, China published a "Catalogue of Encouragement" to guide foreign investment in new and high-tech technologies (issued jointly by the Ministry of Commerce and Ministry of Science and Technology, June 2, 2003). The catalog lists 11 categories grouped into 917 product items. The catalog reflects the priority industries and key technologies identified by China and includes the following sectors: electronics and information, software, air and space industries, optical and electrical integration, biomedical and medical machinery, new materials, new energy and high-efficiency energy conservation, environmental protection, territorial space and ocean, atomic applications, and modern agriculture.

Fiscal Incentives Given by Banks to Encourage R&D and S&T Development

China encouraged banks to provide funds for new and high-tech firms to engage in technology research and development activities. According to a policy issued by China's State Council (Document No. 12, March 1991), banks can issue long-term bonds to raise funds from the public to support the industrial development in the new and high-tech development zones. Relevant government departments are allowed to set up venture capital investment funds, and in some zones venture capital firms are also allowed to set up.

Policy Observations

Juxtaposing international experience with the current state of the innovation system in China and the (fiscal) incentives being offered lays the foundation for a number of suggestions.

First, the growth rate of spending on R&D in China has been unusually rapid. As a percentage of GDP, it almost doubled between 1998 and 2004, and given that the economy was expanding annually at a 9 percent rate, the absolute increase was large. While a further increase of R&D outlay to 2 percent of GDP, and more over the longer term, would be desirable, too rapid an increase risks misallocating resources as the availability of S&T manpower and infrastructure constrain the absorptive capacity for R&D. Many innovations that would strongly influence productivity are unrelated to product or process innovation and depend instead upon innovations in organization and management, as well as the effective assimilation of IT. The big gains registered by the finance, retail, wholesale, and logistics sectors in industrial countries can be traced to such "soft" innovation and are only loosely related to formal R&D. Although China will benefit from R&D, which leads to more patents and good scientific publications, new products, and better production processes, it will gain as much if not more from organizational changes that raise efficiency, including, very importantly, the efficiency of R&D. This might not require adding to fiscal incentives in the near term, as the ones on offer seem to be working, but China would benefit from an acceleration of enterprise reform, for example through the privatization of state-owned manufacturing enterprises (Yusuf, Nabeshima, and Perkins 2005).

Two objectives in this regard would be to (1) refine the statistical estimation of R&D by valuing tax expenditures and (2) introduce other changes that make Chinese statistics comparable to those of the Organisation for Economic Co-operation and Development (OECD). The government could also make a greater effort to evaluate R&D spending and, on the basis of the findings, to improve its quality. This might be coupled to more modest increases in the volume of expenditure on R&D.

Second, international experience recommends some shifting of emphasis from direct budgetary supports of R&D to R&D tax credits, which leave to the firms the decision on what to do research on. There are several variants of such fiscal mechanisms to select from (see table 7.1), and reasonably clear evidence that such incentives have an effect. However, the scale of the effect will depend on the structure and coverage of the corporate tax system. If the effective rates of taxation are relatively low and the coverage narrow, as is the case in China, the leverage that would be exerted by tax credits or depreciation allowances might not be great in the immediate future. Perhaps five years or a decade hence, the impact could increase. Thus, a gradual phasing in of fiscal incentives might be in order to supplement existing mechanisms for promoting R&D.

Third, because the innovation system in China is at the technology absorption stage and experienced research managers are still scarce, the incentives offered to foreign companies to set up R&D facilities can be a potent means of training Chinese researchers, bringing research management skills to China, and inducing the circulation of researchers from abroad. Heterogeneity in the research community is a great asset and needs to be encouraged. It is one of the surest ways of increasing the flow as well as the quality of innovation. However, given how attractive China is to foreign companies wanting to relocate some of their R&D assets, it is not obvious that more fiscal incentives are needed. Stronger protection for intellectual property and efforts that enrich the urban cultural environment might be of greater significance.

Fourth, the substantial incentives to Chinese researchers abroad are working, and as opportunities expand, more and more S&T talent should flow back to China. At this point, there appears to be little need to add to the incentives being offered. The emphasis should be on quality rather than volume, as with R&D spending in general. Attracting back some seasoned "stars" is much more useful. At this stage, allowing others to stay abroad to gain experience and sharpen their skills might be more advantageous for the longer-term development of the innovation system. Ten years from now when China is operating at the frontier of many technologies, the country is more likely to derive benefit from an increase in the flow of returnees.

Fifth, it is desirable to rigorously examine the gains from the multiplication of technology zones and the returns accruing from the fiscal incentives provided (this underscores the point made above with respect to the assessment of R&D spending). If, as some international experience suggests, the benefits in terms of technological advances and productivity are small, then trimming back the fiscal inducements being given would help to strengthen budgetary finances and lessen resource misallocation without limiting the development of the innovation system.

Sixth, while the incentives provided have resulted in a multiplication of state-sponsored venture capital firms, the provision of venture capital remains subject to many administrative requirements that impose significant transaction costs. Removing some of the administrative hurdles would be a positive step. Even more helpful would be measures that increase the participation of experienced foreign venture capitalists and angel investors, who are the ones best positioned to add value through expert advice and contacts. The inexperienced venture capitalists do very little other than provide funding, and that mainly to established firms in safe sectors, rather than high-tech start-ups. Angel investors can have an especially large role; they supplied a quarter of the early-stage financing for innovative start-ups in the United States (Branscomb 2003). For private venture capitalists to take a strong interest, the foreign equity ownership and the use of initial public offerings (IPOs) for liquidating their equity stake or merger and acquisition as an exit venue will have to be institutionally facilitated. This could be a part of a broader reform of the financial system.

Seventh, and finally, because the quality of S&T workers is so critical to the effectiveness of the innovation system, fiscal support, at least for the leading universities, would be a sound investment. A gradual increase in spending on basic research, not just in science but also other fields, would also be desirable. Currently, even China's leading universities are under pressure to diversify sources of funding and pursue commercial options. Up to a point, this is healthy, but if it detracts from the quality of education imparted and the quality and volume of basic research, then the innovation system will be the loser.

Annex: China's Innovation since 1995

Patent Applications and Patents Granted

China's patent applications have increased dramatically over the past decade in terms of both patent applications and patents granted. The number of applications filed (for all three types of patents) have increased from 70,000 in 1996 to 353,800 in 2004; patents granted totaled 190,000 in 2004 (table 7A.1).

China's applications filed for international patents also increased very rapidly. According to reports by the World Intellectual Property Organization (WIPO) of the United Nations for 1978–2004, China fared very well.[4] In 2004 it filed 1,705 invention applications, an increase of 32 percent compared with the previous year, and ranked 14th in international application percentage (1.4 percent). Among developing countries China ranks second in terms of international applications filed.

Invention applications filed in the United States by Chinese residents were up from 695 in 2001 to 1,132 in 2004, and in the same period patents granted by U.S. Patent and Trademark Office were also up from 239 to over 550 (table 7A.2).

Table 7A.1 *Patent Applications Filed with and Granted by the State Intellectual Property Office, 1995–2004*

Reporting year	Applications for patents filed by			Grants of patents to		
	Residents	Nonresidents	Total	Residents	Nonresidents	Total
1995	10,066	31,707	41,773	1,530	1,863	3,393
1996	11,698	41,016	52,714	1,383	1,593	2,976
1997	12,786	48,596	61,382	1,532	1,962	3,494
1998	14,004	68,285	82,289	1,653	3,082	4,735
1999	15,742	73,300	89,042	3,097	4,540	7,637
2000	25,592	96,714	122,306	6,475	6,881	13,356
2001	30,324	118,970	149,294	5,395	10,901	16,296
2002	40,346	140,910	181,256	5,868	15,605	21,473
2003	56,769	48,549	105,318	11,404	25,750	37,154
2004	65,786	64,347	130,133	18,241	31,119	49,360

Sources: Data are from the State Intellectual Property Office of the People's Republic of China and the World Intellectual Property Organization (U.N.) for the years 2003, 2004, and 2005.

Table 7A.2 *Patent Applications Filed by and Granted to Chinese Residents in the U.S. Patent and Trademark Office, FY 2001–FY 2005*

Item	2001	2002	2003	2004	2005
Patent applications filed	695	966	1,230	1,132	2,043
Patents granted	239	347	442	551	583

Source: Data retrieved from U.S. Patent and Trademark Office. http://www.uspto.gov/web/offices/com/annual/2005/2005annualreport.pdf.

China's Scientific Publications

In terms of publications, China improved from 15th in the world in terms of publications listed in Science Citation Index, Engineering Index, and Index to Scientific and Technical Proceedings in 1990 to fifth in 2003. Table 7A.3 shows China's publications catalogued in refereed international journals since 1998.

Table 7A.3 *Publications Catalogued in Refereed Scientific Journals, 1998–2003*

Item	1998	1999	2000	2001	2002	2003
Number of S&T papers catalogued	35,003	46,188	49,678	64,526	77,395	93,352

Source: China National S&T Output Indicators. http://www.sts.org.cn/tjbg/cgylw/documents/2005/051125.htm.

China's Technology-Intensive Exports

The high-tech exports as a part of total manufactured exports have risen very rapidly too: they almost tripled their share from 10 percent of manufactured exports in 1995 to 27 percent in 2003 (table 7A.4).

Table 7A.4 *China's Increasing High-Tech Exports, 1995–2003*

Item	1995	2000	2001	2002	2003
High-technology exports (% of manufactured exports)	10.1	18.6	20.6	23.3	27.1

Source: China National S&T Output Indicators. http://www.sts.org.cn/tjbg/jsmy/documents/2003/05020.htm.

Notes

1. The measures are based on documents (and Web sites) published by the Ministry of Science and Technology, Ministry of Finance, China Tax Administration, Ministry of Commerce, and the Tax policy Regulation on National New and High-Tech Industrial Zones, issued by the State Tax Administration on March 16, 1991. This regulation is still in effect.

2. The project was sponsored by the Scientific Research Foundation for the returned overseas Chinese, Ministry of Education. http://www.moe.gov.cn on.

3. http://www.moe.gov.cn on; http://www.mop.gov.cn.

4. Data available from http://www.wipo.int/ipstats/en/statistics/pct.

References

Bloom, Nick, Rachel Griffith, and John van Reenen. 2002. "Do R&D Tax Credits Work? Evidence from a Panel of Countries 1979–1997." *Journal of Public Economics* 85 (1): 1–31.

Branscomb, Lewis B. 2003. "National Innovation Systems and U.S. Government Policy." Presented at the "International Conference on Innovation in Energy Technologies." September 30.

Cherbonnier, Frederic. 2002. "The Efficiency of the Crédits Recherche Instrument and How It Evolved." Presented at the workshop "The Use of Fiscal Incentives to Boost Innovation," Brussels, April 15–16.

Ciccone, Antonio, and Elias Papaioannou. 2005. "Human Capital, the Structure of Production, and Growth." CEPR Discussion Paper 5354. London: Centre for Economic Policy Research.

Cullen, Julie Berry, and Roger H. Gordon. 2002. "Taxes and Entrepreneurial Activity: Theory and Evidence for the U.S." NBER Working Paper 9015, National Bureau of Economic Research, Cambridge, MA.

David, Paul A., Bronwyn H. Hall, and Andrew A. Toole. 2000. "Is Public R&D a Complement or Substitute for Private R&D? A Review of the Econometric Evidence." *Research Policy* 29 (4–5): 497–529.

García-Quevedo, José. 2004. "Do Public Subsidies Complement Business R&D? A Meta-Analysis of the Econometric Evidence." *Kyklos* 57 (1): 87–102.

Gordon, Robert J. 2004. "Why Was Europe Left at the Station When America's Productivity Locomotive Departed?" NBER Working Paper 10661, National Bureau of Economic Research, Cambridge, MA.

Hall, Bronwyn H., and John van Reenen. 1999. "How Effective Are Fiscal Incentives for R&D? A Review of the Evidence." NBER Working Paper 7098, National Bureau of Economic Research, Cambridge, MA.

Hambling, David. 2005. *Weapons Grade: How Modern Warfare Gave Birth to Our High-Tech World*. New York: Carroll and Graf.

Hervik, Arild. 2002. "Why Did Norway Decide to Introduce R&D Tax Incentives?" Presented at the workshop "The Use of Fiscal Incentives to Boost Innovation," Brussels, April 15–16.

Heytens, Paul, and Harm Zebregs. 2003. "How Fast Can China Grow?" In *China: Competing in the Global Economy*, ed. Wanda Tseng and Markus Rodlauer, 8–29. Washington, DC: International Monetary Fund.

Hezemans, Anja. 2002. "Promoting Business R&D by Fiscal Incentives for R&D Personnel." Presented at workshop "The Use of Fiscal Incentives to Boost Innovation," Brussels, April 15–16.

Hu, Mei-Chih, and John A. Mathews. 2005. "National Innovative Capacity in East Asia." *Research Policy* 34 (9): 1322–49.

Jaffe, Adam B., Josh Lerner, and Scott Stern, eds. 2002. *Innovative Policy and the Economy*, vol. 2. Cambridge, MA: MIT Press.

Jorgenson, Dale W., and Kazuyuki Motohashi. 2005. "Information Technology and the Japanese Economy." NBER Working Paper 11801, National Bureau of Economic Research, Cambridge, MA.

Jorgenson, Dale W., S. Ho Mun, Jon D. Samuels, and Kevin J. Stiroh. 2007. "Industry Origins of the American Productivity Resurgence." Background paper, Harvard University, Cambridge, MA.

Kemppainen, Hannu. 2002. "Some Good Reasons to Abandon R&D Tax Credits: the Case of Finland." Presented at the workshop "The Use of Fiscal Incentives to Boost Innovation," Brussels, April 15–16.

Mamuneas, Theofanis P., and M. Ishaq Nadiri. 1996. "Public R&D Policies and Cost Behavior of the US Manufacturing Industries." *Journal of Public Economics* 63 (1): 57–81.

McKinsey Quarterly. 2002. "The Wal-Mart Effect," January.

Nabeshima, Kaoru. 2004. "Technology Transfer in East Asia: A Survey." In *Global Production Networking and Technological Change in East Asia*, ed. Shahid Yusuf, M. Anjum Altaf, and Kaoru Nabeshima. New York: Oxford University Press.

OECD (Organisation for Economic Co-operation and Development). 2004. *Understanding Economic Growth*. New York: Palgrave Macmillan.

Pedrotti, Tony. 2002. "R&D Activity in the UK: UK Government Activity to Encourage Increased Levels of R&D." Presented at the "Conference on the Use of Fiscal Incentives to Boost Innovation," Brussels, April 15–16.

Rosa, Maria. 2002. "Innovation as the Explicit Target of Spanish Tax Incentives." Presented at the workshop "The Use of Fiscal Incentives to Boost Innovation," Brussels, April 15–16.

Russo, Benjamin. 2004. "A Cost-Benefit Analysis of R&D Tax Incentives." *Canadian Journal of Economics* 37 (2): 313–35.

Santos, Rui. 2002. "The Role of Fiscal Incentives in an Integrated Modernisation Policy." Presented at the workshop "The Use of Fiscal Incentives to Boost Innovation," Brussels, April 15–16.

Scotchmer, Suzanne 2004. *Innovation and Incentives*. Cambridge, MA: MIT Press.

Skinner, Jonathan, and Douglas Staiger. 2005. "Technology Adoption from Hybrid Corn to Beta Blockers." NBER Working Paper 11251, National Bureau of Economic Research, Cambridge, MA.

Steil, Benn, David G. Victor, and Richard R. Nelson, eds. 2002. *Technological Innovation and Economic Performance*. Princeton, NJ: Princeton University Press.

Temple, Jonathan. 1999. "The New Growth Evidence." *Journal of Economic Literature* 37 (1): 112–56.

———. 2001. "Growth Effects of Education and Social Capital on OECD Countries." Discussion Paper 2875, Center for Economic Policy Research, Washington, DC.

UNCTAD (United Nations Conference on Trade and Development). 2005. *World Investment Report 2005: Transnational Corporations and the Internationalization of R&D*. New York: United Nations.

van Dalen, Hendrik P., and Arjo Klamer. 2005. "Is Science a Case of Wasteful Competition?" *Kyklos* 58 (3): 395–414.

van Pottelsberghe de la Potterie, Bruno. 2008. "Europe's R&D: Missing the Wrong Targets?" Bruegel Policy Brief 2008/03, Bruegel, Brussels.

Wallsten, Scott. 2004. "Do Science Parks Generate Regional Economic Growth? An Empirical Analysis of Their Effects on Job Growth and Venture Capital." Working Paper 04-04, AEI-Brookings Joint Center for Regulatory Studies, Washington, DC.

Wieser, Robert. 2005. "Research and Development Productivity and Spillovers: Empirical Evidence at the Firm Level." *Journal of Economic Surveys* 19 (4): 587–621.

Yusuf, Shahid, Kaoru Nabeshima, and Dwight H. Perkins. 2005. *Under New Ownership: Privatizing China's State-Owned Enterprises*. Stanford, CA: Stanford University Press.

Zhao, Min. 2005. "Policies for the 11th Foreign Capital Utilization Plan." Background paper, World Bank, Beijing.

8

China's Tax Policies for Promoting Innovation

Yaobin Shi

International competition is driven primarily by the level of scientific and technological development in each country. Globalization and international capital flows have increased China's integration with the world economy. Although China's economy benefits greatly from globalization and latecomer advantages, it will also face new challenges because of the acceleration of scientific and technological progress as well as the further strengthening of comparative advantages in industrialized countries. The experiences of many countries show that science and technology (S&T) are major drivers of production and help determine a nation's political status and overall strengths. Promotion of scientific and technological innovation is, therefore, a vital part of a country's strategy. To this end, governments often adopt a multitude of public policies, including taxation policies, in order to promote scientific and technological innovation and economic growth.

To ensure China's sustained growth, achieve the government's target of doubling the country's GNP by 2020, and establish a well-off society in an all-round way, China must work toward the following: adjust its economic structure and growth patterns; follow the new industrialization road and construct a saving-type society, and reduce consumption and waste of resources and energy by relying on scientific technology development and progress. Enhancing the capacity of firms to innovate and promoting their technological progress are keys steps for China to enhance the competitiveness of its traditional industries. Doing so is also necessary for the country to optimize its industrial structure, transform economic growth patterns, and ensure the success of its long-term development strategy.

In light of numerous government statements on innovation for development, China should use its tax system and tax policies to promote scientific and technological innovation.[1] Combining both theory and practice, this chapter analyzes China's existing tax policies that support innovation and suggests several next steps for implementing more innovation-supporting tax policies and systems.

This chapter begins by analyzing the theoretical basis for the government to formulate tax policies to promote technological innovation, using the concept and characteristics of innovation as a starting point. It then looks at the challenges and opportunities faced by China regarding technological innovation and advocates that the government should adopt effective tax policies to promote technological innovation. In the last parts, it introduces China's existing tax policy incentives for technological innovation, assesses several problems in the current policy, and suggests adjustments in the tax regime to improve innovation.

Autonomous innovation is necessary for countries—developing countries in particular—to improve their comparative advantages. However, innovation carries a large degree of uncertainty, so public policy is essential to fostering innovation. Compared with monetary policies, fiscal and tax policy has unique advantages on economic structure adjustment that can encourage investments in science and technology, promoting indigenous innovation. Consequently, it is one of the most effective policy instruments available. Nevertheless, formulating tax-based innovation policies—given China's unique situation and constraints—involves important theoretical and practical issues. This chapter is written specifically to address issues related to harnessing the unique advantages of tax policies for promoting indigenous innovation among enterprises in China.

The Theoretical Basis for Government Tax Policies Designed to Promote Innovation

Innovation first appeared as a concept in *The Theory of Economic Development*, by economist Joseph Schumpeter (1912/1990). According to Schumpeter, technological innovation introduces new inventions and discoveries into the production system, which in turn generates shock effects. He emphasized that economic development is an evolutionary process with technological innovation at its core and firms as the main driving force behind innovation. Furthermore, the success or failure of innovative activities depends on the social and economic context in which they operate. Schumpeter's viewpoints provide useful insights into technological innovation, policies, and theories.

Since the 1960s, Western economists such as Kenneth Arrow, Richard R. Nelson, and David Romer have applied the theoretical hypothesis of market failure to the policies for technological innovation. They concluded that a market failure in the process of technological innovation exists, and this provides the rationale for government intervention. Information asymmetries, nonexclusivity, risk, and uncertainty cause private benefits to innovators to be lower than societal benefits. This in turn weakens enthusiasm and incentives for firms to innovate and ultimately affects the innovative capacity of both firms and society as a whole. Thus, the presence of positive externalities and market failure give reason for government intervention.

Tax policy is one of the tools used most frequently by governments. The primary function of taxation is to pool revenue for the treasury, but at the same time it is also a macroeconomic instrument used by the government to promote scientific and technological innovation. A standardized, uniform tax system can promote fair competition for various market players and provide strong fiscal support for innovation and the development of public services. Government can also use discretionary tax policies to assist firms in innovation by reducing investment costs and raising returns on investment. Tax policy can also guide firms to undertake more research and development (R&D). Alternatively, limited tax expenditures can induce social funds into areas for innovative investments.

The experiences of many countries show that taxation policies play an important role in encouraging enterprise innovation by affecting several factors, such as levels of R&D, accumulation of human capital, and entrepreneurial behavior.

Why Improving Tax Policies for Innovation Is Necessary for Scientific and Technological Development in China

Economic growth in developed countries over the past two centuries has gone through the stages of factor accumulation, intensive management, and knowledge (including technological) innovation. In particular, technological innovation has promoted a leapfrog approach to development and economic growth. Scientific and technological innovation is an important component of many countries' national strategies to increase their capabilities and foster strong competitive advantages. Twenty countries ranked high in terms of innovation—including the United States, Japan, and Finland, and all exhibit similar characteristics. For example, rates of contributions by science and technology to GDP growth are over 70 percent, and R&D inputs as a percentage of GDP are generally over 2 percent. Additionally, the degree of dependence on external technologies in these countries is generally lower than 30 percent, and the combined number of patents account for 99 percent of the global total (Fiscal Science Research Institute 2007).

Although enterprises are the main forces of technological innovation, it is also a joint undertaking by the government and enterprises. Governments in developed countries tend to play a strong guiding role in the production, dissemination, and application of scientific and technological knowledge. Public policies in these countries, including preferential tax treatment, are used to increase inputs into science and technology and to promote indigenous innovation.

Over the past few years, China has made great progress in scientific and technological innovation; however, it still lags behind many advanced nations in terms of overall scientific and technological development. This is owing to low levels of original and integrated innovation at the local level, ineffectiveness of the strategy of "exchanging technologies with market," and weak indigenous innovation capabilities among firms. These are major problems affecting China's industrial competitiveness and also hindering the sustained growth of its economy.

The *World Competitiveness Yearbook 2005*, released by the Swiss-based IMD (2005), shows that in 2004 China ranked 24th among 49 major countries in terms of scientific and technological innovation capacity. According to the report, scientific and technological progress contributed to 39 percent of China's economic growth, while the degree of dependence on external technologies by firms was as high as 50 percent. In basic research, only 15 percent of China's academic fields meet internationally advanced levels. In industrial technologies, China's patents for new inventions are only 1/30 those of Japan and the United States and 1/4 of the Republic of Korea's. R&D expenditure in China accounts for 1.3 percent of GDP, and the ratio of investments in R&D to investments for commercialization of R&D results is below 1:5. On average, only 10–15 percent of scientific and technological discoveries in China are commercialized. China, on an international level, is still at the low end of the U-curve with an industrial profit rate of merely 5 percent.

Given the large gap in levels of innovation between China and advanced nations, China faces many challenges in its economic and social development. In order to keep up with increasing fierce international competition, China needs to increase its competitiveness through self-owned intellectual property rights and the optimization of its industrial structure. Scientific and technological development is vital for sustaining China's economic growth and increasing the efficiency of resource consumption. Public science and technology is also important for national security, social stability, and development of China's rural regions.

The 5th Plenum of the 16th Central Committee of the Communist Party of China pointed out that it is imperative for China to "regard the strengthening of indigenous innovation capabilities as the strategic basis for scientific and technological development and the central thrust for adjusting of industrial structures and transforming economic growth patterns; and towards this end, exerting great efforts to increase indigenous and integrated innovation capabilities and the capabilities of introduction, digestion, absorption, and re-innovation." The "Outlined National Program for Medium- and Long-Term Development of Science and Technology (2006–2020)" clearly stipulates that the guiding principle for China's scientific and technological development over the next 15 years is to foster indigenous innovation and support the country's development (Ministry of Science and Technology 2006). The plan calls for an increase in R&D inputs nationwide to over 2.5 percent of GDP by 2020 and stipulates the rate of contribution of scientific and technological progress to GDP to reach over 60 percent. Also, the plan stipulates the goal of a reduction in the dependence on external technologies to under 30 percent and an increase in the numbers of patents for new inventions and internationally quoted science papers to the level of the world's top five.

These goals pose challenges for China's tax policy, which plays an important role in fostering innovation. The tax regime must be long term, flexible, and supportive of indigenous innovation by firms. It must also help create a favorable policy environment for the conversion of scientific and technological innovation into actual production capacities.

China's Existing Tax Incentives for Scientific and Technological Innovation

China has been enhancing its capacity in scientific and technological innovation, particularly after 1978. In the past decades, tax-related policies played an important role in promoting innovation. It has been observed that new issues have to be considered with regard to the economic growth to meet the changing economic environment.

Overview of Existing Tax-Related Policies for Promoting Innovation

Since the reform of its tax system in 1994, China has implemented a series of tax policies to promote innovation. These policies were designed to aid the development of China's new high-tech industries, encourage enterprise R&D and dissemination of scientific technology results, and provide incentives for S&T personnel (see box 8.1). These policies cover almost all of China's existing tax categories and come in various forms: indirect forms such as pretax deductions, accelerated depreciation, and tax credit for investment; or direct forms such as tax reductions or exemptions. Direct tax reductions and exemptions are the most commonly used policy tools. Many of these tax incentives, such as those for new and high-tech industries, have been in effect since their introduction; many have been in force for over 20 years. Data, though incomplete, indicate that tax reductions and exemptions for new and high-tech firms amounted to 25 billion yuan (Y) in 2003. In 2004, total tax reductions and exemptions for science and technology exceeded Y 70 billion. These tax incentives have helped to create a policy environment that is conducive to innovation and to achieving the government's objective of sustaining the country's macroeconomic development.

Problems Existing in the Current Tax Preferential Policies for Innovation

Studies show that China's current system of tax policies to encourage innovation has several problems that require attention. First, most of China's tax policies apply only to results of innovation instead of the process of innovation. Only when an enterprise is recognized as a high-tech firm or when R&D has led to the development of scientific and technological results can the enterprise enjoy the relevant tax benefits. During the actual R&D process and experimentation, tax preferences for firms are often insufficient.

Box 8.1 *Examples of Tax-Based Innovation Policies in China*

1. Tax policies for promoting the development of high-tech firms. High-tech firms newly established within national high-tech industrial development zones are exempt from income tax until two years after the start of production (in 2006, this provision was amended to "year of profitability" instead of "start of production"). After the two years, these firms enjoy a lower corporate income tax rate of 15 percent.

2. Tax policies for promoting technological research. Firms that see an annual increase in R&D expenditures of over 10 percent are eligible to receive an additional deduction on their taxable income up to 50 percent of the actual cost of expenditures. Firms that did not deduct all the R&D expenditures could carry over their remaining expenditures so that they can claim deductions in the following year. In 2006, the 10 percent restriction was removed, and firms could claim 150 percent of the actual expenditures on R&D as tax deductions for the same year. For instruments and equipment used for R&D, their value can be included as expenses for claiming tax deductions if the value was lower than Y 300,000. If the value exceeded Y 300,000, the assets would instead be subject to accelerated depreciation benefits.

3. Tax policies for the dissemination of scientific and technological results. Scientific institutes, institutes of higher education, and certain types of vocational schools enjoy exemptions from corporate income tax for their income derived from technological services such as technical training or consulting services. Businesses and private individuals engaged in similar activities are exempt from the business tax. Foreign enterprises can also receive a lower corporate income tax rate of 10 percent—if approved by tax authorities under the State Council—for royalties earned by providing technical know-how for scientific research and the development of key technologies. Foreign enterprises can also be exempt from income taxes if they have very advanced technologies or other advantages

4. Tax incentives for encouraging S&T personnel and human capital. Bonuses and prizes awarded by government or international organizations for S&T research and development are exempt from the personal income tax. In addition, commissions received by private individuals engaged in technology transfer and labor services can also be deducted from their personal income tax, provided that valid proof of the expense can be presented. For staff working on software and circuit development, their salaries can be fully deducted for corporate income tax too.

5. Accredited enterprise technology centers. When accredited centers import apparatuses, instruments, chemical agents, or technical documents, which are used for the development of new technologies and are not available in China, these imported items are exempt from import duties and import-stage value added taxes.

Source: Author's compilation.

Second, China's tax incentives are disorganized and inconsistent. For example, tax preferences for S&T-based firms are applied in connection with the value added tax (VAT) and the import-stage tax—a practice that hurts efforts to standardize the VAT and also creates problems such as substituting public investment for S&T with taxes. In contrast, some firms do not receive any tax preferences at all, even though by internationally accepted best practices they should. So far there are no specific tax preferential policies for national S&T parks and incubators or small and medium enterprises (SMEs), which form the backbone of indigenous innovation.

Third, provisions on pretax deductions are inadequate. Deductions for employee education expenses are too low, and for employees in many firms, wages and salaries do not receive the full deduction. These problems have restricted the development of human capital in many high-tech industries.

Fourth, some tax incentives have undesirable effects. Tax benefits for foreign-invested enterprises, for example, because of jurisdictional issues often become additional revenue accrued by other countries instead of benefiting the firms. Other firms that are subject to the assessment tax collection method often find it difficult to benefit from tax incentives. Additionally, tax incentives for high-tech firms are different depending on the firm's location—firms in high-tech development zones receive different treatment from those registered outside of these zones. The difference in treatment harms competition and opens the door to loopholes in tax collection and administration.

Fifth, tax incentive policies are often not coherent and their legal status is relatively low. Local governments often unveil tax incentives in the form of tax rebates without authorization. This practice weakens the role of tax policies in fostering innovation. In light of these problems, improving China's tax system to better promote indigenous scientific and technological innovation entails three main objectives: making full use of existing relevant tax policies, making gradual improvements in the existing tax incentives, and harnessing the full potential of tax policies to promote innovation.

Tax Policy Options in Support of Scientific and Technological Innovation

It is hard to conclude that science and technology contributed less to China's economic growth in the past years, as some surveys reported recently. To make economic development more sustainable, tax-related policies that support scientific and technological innovation are being considered.

The Rationale for Using Tax Policies to Promote Innovation

The key to enhancing China's innovation capacity is to strengthen indigenous innovation through enterprises. It is important to keep in mind that taxation is not the major constraint on firms' ability to innovate. Rather, tax policies are inducers designed to create a favorable tax environment for firm innovation. To encourage firms to innovate, issues such as fundraising systems, capital market provisions, entry and exit mechanisms, and so forth have to be addressed comprehensively. Costs such as new equipment, technology, and R&D expenditures are necessary for firms to increase their competitiveness and achieve higher returns on investments. Market mechanisms should be allowed to work freely, and the government should play a guiding role rather than replace firms in the decision-making process. Strengthening indigenous innovation through enhancing the capabilities of firms, therefore, should be the government's primary objective.

Using international experiences as reference to reform China's tax regime, research and theory suggest that technological innovation should coincide with institutional innovation in order to increase efficiency and effectiveness. The implementation of existing innovation-related tax policies needs to be improved in order to realize the full benefits of those policies. Given this, it is also essential to transform the government's mentality toward the role of tax policies and encourage policy innovations in designing key tax incentives.

Principled Requirements for Using Taxation to Support Innovation

Tax policies designed to promote innovation should be compatible with the overall direction and implementation of China's tax system reforms. In line with these objectives, the government should accelerate the reform of the VAT and prepare for the implementation of the new income tax law. Reforming the VAT will eliminate double taxation, encourage investment, and assist in the transformation of China's industrial structure by improving firms' technology level and competitiveness.

The new corporate income tax law will streamline pretax deductions and make much-needed changes in the tax rates and reductions or exemptions for both domestic and foreign-invested enterprises. For instance, under the new law, tax incentives will favor new and high-tech firms, promotion of R&D activities, energy and resource conservation, and environmental protection. These changes are designed to reflect China's new industrial policy and to create a level playing field for all kinds of enterprises. Innovation is largely driven by capital or technology-intensive firms, so the new policies should appropriately reflect these firms as the main beneficiaries of the tax reforms. Thus, tax measures promoting scientific and technological development must be coordinated with the overall objectives of China's tax system so that they do not hinder or become an obstacle to the government's tax reforms.

The government must also accurately define the boundaries between financial support and tax incentives to avoid abuse of tax policies. Fiscal expenditure should mainly be used for basic research, high technologies with strategic importance and key common technologies, and other quasi-public goods that promote sustainable development, such as environmental protection and energy conservation. In contrast, tax preferences should focus on encouraging and inducing enterprises' indigenous innovation through incentives measures such as tax deductions, exemptions, or accelerated depreciation. The government should use neutral tax types such as the VAT and the business tax as little as possible to promote innovation, and tax reform efforts for innovation should be uniform, transparent, and effective.

Tax policies to support innovation should be compatible with internationally accepted practices and World Trade Organization rules. Preferential tax policies should avoid influencing firms' decision making on product development. The government needs to follow generally accepted practices and place priority on supporting key common technologies and technologies with strategic importance and on increasing the competitiveness of China's key industries. It is also important to fully utilize fiscal approaches such as subsidies to promote indigenous innovation among firms.

Options for Developing an Innovative Tax System

Option 1   Reform main tax categories and enhance the catalyst role of tax policies in technological innovation.

1. Increase efforts to transform the VAT system. The main objective is to shift the VAT away from production-based taxes and toward consumption-based taxes and allow firms to deduct the input tax of fixed assets from the output tax. These changes will alleviate the tax burden for many firms, encourage them to make more investments in innovation, and help them upgrade their technologies and equipment. An ongoing pilot project in northeast China that

is experimenting with VAT reforms shows how changes in the VAT are transforming China's old industrial bases and serves as a prototype for VAT reforms on the national level.

2. Create sound conditions for the implementation of the new income tax laws. Implementation of the laws should take into account indigenous innovation among firms and ways to upgrade the overall industrial structure. The corporate income tax regime needs to clearly define provisions regarding the compensations for firms' capital, labor, and technology expenses in order to establish a fair and reasonable tax system and create a uniform, open, and level playing field for all enterprises. Adjustments also need to be made in existing income tax incentives and pretax deduction items and standards in order to encourage enterprise R&D, autonomous innovation, and the conversion of scientific results into new products.

Option 2 Streamline and improve existing preferential tax policies for innovation.

1. Improve the design of tax preferences to better promote fair competition and technological innovation among enterprises. Preferential tax policies are essentially waivers of certain taxes on enterprise revenues by the government, but such waivers come in various forms and generate different effects. In comparison with international experience, China should shift its focus from direct preferential treatments toward indirect incentives and use a variety of methods such as accelerated depreciation, tax credit for certain investments, compensation for losses, and deductions for expenses. The government should also offer firms a 150 percent pretax deduction for R&D expenditure. Accelerated depreciation measures such as double remaining-balance deductions should be applied to instruments and equipment used in R&D activities if their value exceeds the government-set limit. The government should also further streamline income tax credit policies on equipment investments. The implementation of the above measures will further reduce costs associated with innovation and speed up the turnover and recovery of funds incurred toward technological innovation.

2. Further improve the tax incentives for new and high-tech enterprises. High-tech enterprises are the foundation of technological innovation, so it is essential to improve and expand existing preferential

tax policies for high-tech enterprises. New tax incentives should also be introduced for modern service industries that employ science and technology, such as software development, cartoons and animation, and information services.

3. Use tax incentives to provide enterprises with additional funds for conducting innovation-related activities. In addition to reducing costs associated with innovation and increasing returns on investments, tax preferences can also channel public funds into small and medium high-tech enterprises, alleviating financial difficulties for these firms. For example, a new tax benefits policy on venture investments allows for 70 percent of the capital invested in the high-tech SMEs to be deducted from taxable income. Efforts should also be made to regularize tax benefits for grants and donations given to technological innovation funds assisting high-tech SMEs and other funds to promote indigenous innovation.

4. Apply preferential tax policies to strengthen human capital development in science and technology. It is essential to change the traditional perception that physical resources are more important than human resources, and instead apply a "people first" approach in designing preferential tax policies. It is also important to implement, borrowing from international experiences, a set of preferential tax policies that are integrated with the national human capital strategy, to increase the efficiency of labor productivity and the country's overall talent pool. Relevant income tax incentives for high-tech talents should be strengthened and other measures, such as pretax deduction of educational expenses, should also be gradually improved.

5. Use preferential tax policies to encourage the conversion of scientific and technological results into products. Taxation plays an important role in helping to commercialize the results obtained from scientific and technological innovation. Improvements in tax policies would encourage scientific research institutes and institutions of higher learning to work together with enterprises in conducting R&D activities—helping research institutes to operate in a more market-oriented way—and promote the dissemination and use of scientific and technological results in production activities. Tax incentives should also be applied to S&T intermediary institutions such as university-based national science and technology parks and incubators of new technology ventures—service providers dedicated to

fostering new high-tech SMEs. They help new start-ups by offering physical space and infrastructure to reduce entrepreneurial risks and costs and increase start-ups' chances for success. Therefore, it is very important to learn from international experiences and to formulate and implement preferential tax policies for S&T parks and incubators.

Note

1. See "Decision on Several Issues in Relation to the Improvement of the Socialist Market Economic System," adopted by the 3rd Plenum of the 16th Central Committee of the CPC; the "Recommendations in Relation to the Formulation of the 11th Five-Year Plan on the National Economic and Social Development of the CPC Central Committee," and the Ministry of Science and Technology's "Outlined National Program for Medium- and Long-Term Development of Science and Technology (2006–2020)."

References

Fiscal Science Research Institute. 2007. "Fiscal Policies in Innovation-Based Nations and Implications for China." Report No. 40, Ministry of Finance.

Government of China. 2006. "Outlines of the People's Republic of China for the 11th Five-Year Plan for the National Economic and Social Development." http://www.gov.cn.

IMD. 2005. *World Competitiveness Yearbook 2005*. Lausanne, Switzerland: IMD International. http://www.imd.ch.

Ministry of Science and Technology. 2006. "Outlined National Program for Medium- and Long-Term Development of Science and Technology (2006–2020)." [In Chinese.] http://www.gov.cn.

Ni Hongri. 2007. "Analysis of Tax Policy and Institutional Improvements Designed to Encourage Indigenous Innovation." [In Chinese.] *Journal of Tax Studies* 1: 6–10.

Qu Shunlan and Lu Chuncheng. 2007. "Indigenous Innovation and Effects of Fiscal and Tax Policies." [In Chinese.] *Journal of Tax Studies* 1: 17–20.

Schumpeter, Joseph. 1912/1990. *The Theory of Economic Development: An Inquiry into Profits, Capital, Credit, Interest, and the Business Cycle*. Trans. Redvers Opie. Hong Kong, China: Commercial Press.

Wang Chunfa. 1998. *Technological Innovation Policy: Theoretical Basis and Instrument Options*. Beijing: Economic Science Press.

Yu Yongding and Li Xiangyang. 2002. *Development Trends of Economic Globalization and the World Economy*. [In Chinese.] Beijing: Social Science Literature Press.

Index

ECO-AUDIT
Environmental Benefits Statement

The World Bank is committed to preserving endangered forests and natural resources. The Office of the Publisher has chosen to print *Innovation for Development and the Role of Government* on recycled paper with 30 percent postconsumer fiber in accordance with the recommended standards for paper usage set by the Green Press Initiative, a nonprofit program supporting publishers in using fiber that is not sourced from endangered forests. For more information, visit www.greenpressinitiative.org.

Saved:
- 7 trees
- 5 million Btu of total energy
- 585 lb. of net greenhouse gases
- 2,428 gal. of waste water
- 312 lb. of solid waste